Chemicals Based on the Light Fraction of Natural Mercaptans

Chemicals Based on the Light Fraction of Natural Mercaptans

V. I. Latyuk
V. I. Kelarev
V. N. Koshelev
K. D. Korenev
and
G. E. Zaikov

Nova Science Publishers, Inc.
New York

For permission to use material from this book please contact us:
Telephone 631-231-7269; Fax 631-231-8175
Web Site: http://www.novapublishers.com

LIBRARY OF CONGRESS CATALOGING-IN-PUBLICATION DATA

Chemicals based on the light fraction of natural mercaptans / K.D. Korenev, V.I. Latyuk and V.I. Kelarev, [editors].
p. cm.
Includes index.
ISBN-13: 978-1-60021-650-3 (hardcover)
ISBN-10: 1-60021-650-1 (hardcover)
1. Thiols--Analysis. 2. Chemical tests and reagents. I. Korenev, K. D. II. Latyuk, V. I. III. Kelarev, V. I.
QD341.S3C485 2006
665.5'38--dc22 2007008431

Published by Nova Science Publishers, Inc. ✢ New York

CONTENTS

INTRODUCTION

Many major oil-and-gas fields in USA, Canada, Russia and other countries keep sulfur-, oxygen- and nitrogen-containing chemicals along with general hydrocarbon mass. These impurities complicate the processing of natural raw hydrocarbons material to fuels and chemicals contaminating them and decreasing their consuming properties. In particular, it is generally known that sulfur-containing impurities lower the stability of hydrocarbon fuels towards oxidation. At the same time these impurities are the cause of corrosive action of the fuels.

Along with other sulfur-containing chemicals, mercaptans (sulfur-substituted analogues of alcohols) are often present at sulfurous oils and gas condensates in noticeable amount. Mercaptans could be extracted with water solutions of sodium and potassium hydroxides. The amount of natural mercaptans that can be potentially isolated this way is dozens of thousand tones per year. The example is the processing of gas condensate from Orenburg (Russia), Crossfield and Line Creek (Canada), Lac (France) and many other fields. However the qualified way of utilization of these mercaptans is not found so far. The most common way today is to burn them with formation of noxious sulfur oxides. Moreover, limited amount of mercaptans is used as odorant (substance possessing strong objectionable odor) for fuel gas.

The present book describes scientific basis for isolation and possible utilization of the light fraction (C_2-C_5) of mercaptans, containing in gas condensate from Orenburg (Russia).

Chapter 1

PROPERTIES AND APPLICATION OF MERCAPTANS (THIOLS) IN ORGANIC SYNTHESIS (REVIEW)

1.1. MOLECULAR STRUCTURE AND PHYSICAL PROPERTIES OF MERCAPTANS (THIOLS)

Mercaptans (thioalcohols) was known as far back as in the Middle Ages to form characteristic sparingly soluble mercury salts (*corpus mercurio aptum*), which gave the name to this class of organic compounds [1]. Mercaptans have a pungent objectionable odor. Being very strong the odor allows detecting the very small amount of C_1-C_5 homologues in the air ($\leq 2{\cdot}10^{-4}$ mg/l) [2]. Mercaptans, especially C_1-thiol, negatively affect the nervous system of humans and animals. Mercaptans are considered as highly-toxic substances, lethal concentration in air is 200-300 mg/l with exposition time of 1 min, admissible concentration limit is 0.9 mg/l [2, 3].

According to the nomenclature rules adopted by the International Union of Pure and Applied Chemistry (IUPAC) the name "mercaptan" was substituted by "thiol" that reflected better the chemical nature of these substances [4]. Nevertheless, the term "mercaptan" is employed widely in certain technical areas [5-8]. The name "thiols" point out their analogy with alcanols [9, 10]. The similarity of many of the properties of thiols and alcohols is determined by the resemblance of the electron distribution at S and O atoms. At the same time atom S has 5 vacant orbitals to bond with other atoms, while atom O has only 4 orbitals able to accept 8 electrons from other atoms [11], so that is the reason of known difference between properties of substances compared.

C-S-H bond angle in C_1-thiol is 100.3 ±0.2° [9] or 96.5° [10] that is close to 90°. That indicates the formation of S-H bond via the overlap of 3p orbital of S atom and 1s orbital of H atom. In methanol molecule this angle is equal to 108.9±2° [9] being close to tetrahedral one (109.3°). Therefore, sp^3-hybridization of oxygen atom orbitals takes place when forming O-H bond [11]. The nature of S-H and O-H bonds in thiols and alcohols correspondingly is responsible for their different strength. The lower S-H bond energy (339 kJ/mole) as compared with O-H bond energy (462 kJ/mole) [10] is one of the reasons for increased activity of thiols regarding to alcohols towards dissociation reactions including thiyl radicals formation [12, 13].

Because of the lower electronegativity of S atom and, consequently, S-H bond thiols possess weaker proton donor ability when forming intermolecular H-H bonds than alcohols [14]. Thus, boiling points of thiols, for example methanethiol and ethanethiol, are 41-59°C lower in comparison with methanol and ethanol [9]. Moreover these thiols are poorly soluble in water [15]. For example only 1.5 g of methanethiol dissolves in 100 ml of water at 20°C. Increasing the molecular mass of thiols, their solubility in water still more decreases. The temperature increase has almost no effect on the solubility. Unlike thiols, methanol and other lower alcohols are blended with water in any proportions [1].

Other polar solvents (alcohols, acetone, nitromethane etc.) dissolve thiols well at ambient and lowered temperatures [15]. Despite the poor solubility, lower thiols undergo acidic dissociation in water [9, 15]. Dissociation constants of C_2-C_6 thiols with different structure lie in the range $2.3 \cdot 10^{-12}$–$3.2 \cdot 10^{-14}$, the same for corresponding alcohols are 10^{-16}–10^{-20}. At the same time alcohol's anions (alcoholates) are stronger bases than thiolates (mercaptides) [10].

Thiols are readily adsorbed at the solid surfaces containing Si and Al oxides. Their adsorption activity is only a little lower than the activity of aliphatic and cyclic sulfides [15]. Intermolecular interaction of SH groups of thiols with surface HO groups of silica and electron-acceptor acid sites of γ-Al_2O_3 plays the governing role in adsorption process [16].

1.2. Reactions of Thiols

First of all, thiols readily react with bases in aqueous and aqueous-alcoholic media forming ionized salts of sodium, potassium and other metals (or nitrogen bases) [9, 10, 15, 17]:

$$RSH + M(OH) \longrightarrow RS^-M^+ + H_2O \qquad (1.1);$$

$$RSH + H_2NC_2H_4ONa \longrightarrow RSNa + H_2NC_2H_4OH \qquad (1.2).$$

In contrast to thiolates, alcoholates are formed by reaction of alcohols with alkali and alkaline-earth metals only [1].

Thiols interaction with organic and inorganic salts of Ag, Cd, Hg results in formation of corresponding non-soluble thiolates [15, 18].

Thiolates of metals (or nitrogen bases) react readily with alkyl halides forming sulfides (thioethers) [9, 15]:

$$RS^-M^+ + HalR' \longrightarrow RSR' + M^+Hal^- \qquad (1.3).$$

Thiolates are noticed to have higher reaction ability in this reaction than alcoholates have.

Mobility of proton in thiol and alcohol molecules allows them to undergo quick esterification by different reagents catalyzing both by acids and by bases [1, 19, 20]:

$$RSH + R'COOH \longrightarrow R'C(O)SR + H_2O \quad (1.4),$$

$$RSH + H_2C\underset{O}{\diagdown\!\diagup}CH_2 \longrightarrow RSCH_2CH_2OH \quad (1.5),$$

$$RSH + H_2C\underset{O}{\diagdown\!\diagup}CHCH_2Cl \longrightarrow RSCH_2CH(OH)CH_2Cl \quad (1.6).$$

Ester-like products, in particular thiochlorocarbonates, are formed in the reaction of thiols with phosgene in the presence of HCl acceptor [7]:

$$RSH + COCl_2 \xrightarrow[-\,HCl]{} RSC(O)Cl \quad (1.7).$$

Under simultaneous treatment by halogen acids and aldehydes thiols convert into corresponding α-halogenoalkylthioethers [21, 22]:

$$RSH + R'CHO + HBr \longrightarrow RSC(R')Br + H_2O \quad (1.8),$$

including alkylthiomethylhalogen derivatives [23]:

$$RSH + CH_2O + HCl \longrightarrow RSCH_2Cl + H_2O \quad (1.9).$$

Another way of synthesis of these derivatives is proposed in [24] consisting in thiol treatment with a mixture of paraform and trimethylchlorosilane at t = 0 – (-5)°C:

$$RSH + 1/n(CH_2O)_n + 2ClSi(CH_3)_3 \rightarrow$$
$$\rightarrow RSCH_2Cl + (CH_3)_3SiOSi(CH_3)_3 + HCl \quad (1.10).$$

The reaction of oxidative sulfochlorination of thiols is described in [25]. This reaction results in alkylsulfochlorides with 45-90% yield:

$$RSH + Cl_2 + O_2 \longrightarrow RSO_2Cl + HCl \quad (1.11).$$

Interaction of thiols only with formaldehyde gives unstable semi-mercaptals and mercaptals [26]:

$$RSH + CH_2O \longrightarrow RSCH_2OH;$$
$$2RSH + CH_2O \longrightarrow (RS)_2CH_2 + H_2O \quad (1.12),$$

while the interaction with ketones results in similar dithio derivatives [27]:

$$2RSH + R^1C(O)R^2 \longrightarrow R^1C(SR)_2R^2 + H_2O \quad (1.13).$$

Thioalkylation of nitromethane [28], indol, antipyrine, β-naphtol [29] and a number of amines [30] is also described.

Simultaneous interaction of thiols with formaldehyde and ketones named the reaction of ketones thiomethylation was carried out for the first time in the presence of sodium hydroxide [31] and sulfuric acid [32]:

$$RSH + CH_2O + R^1C(O)CH_2H \xrightarrow[\text{or } H_2SO_4]{\text{NaOH}} R^1C(O)CH_2SR + H_2O \quad (1.14).$$

This reaction quickly runs at 20-70°C in aqueous-alcoholic medium or in tetrahydrofuran medium with intermediate formation of alkylmethanols ($RSCH_2OH$) under conditions of base catalysis and formation of carbonium-sulfonium cations ($H_2C^+\text{-}SR \leftrightarrow H_2C = SR$) in the case of acid catalysis. Thioalkylation is most likely to run through the mechanism determined for the commonly known Mannich reaction [33].

γ-Ketosulfides (γ-oxoalkylsulfides with alkyl group consisting of two C atoms) producing by these reactions can be also synthesized by thiols interaction with β-halogeno derivatives of ketones [34]:

$$RSH + R'C(O)CH_2CH_2Hal \longrightarrow R'C(O)CH_2CH_2SR + HHal \quad (1.15).$$

or with bis-β-dialkylaminoketones (ketone Mannich bases) [35]:

$$RSH + R^1\underset{\underset{O}{\|}}{C}CH_2CH_2NR^2R^3 \longrightarrow R^1\underset{\underset{O}{\|}}{C}CH_2CH_2SR + NHR^2R^3 \quad (1.16).$$

An important group of reactions is the addition of thiols to unsaturated compounds that is referred as thiylation reaction [12, 13, 15, 36]. In absence of acid catalyst this reaction runs with low yield at elevated temperatures (150-300°C). At the same time, under the influence of, for example, H_2SO_4 thiols react with alkenes, most readily with isoalkenes, already at 46-60°C. Aluminum chloride, sulfides of W, Mo, Ni etc. are proposed to be the acid catalysts for thiylation of alkenes. Under these conditions alkenes add thiols in accordance with Markovnikov rule. The possibility of further synthesis of products of electrophilic addition of thiols to regular and isoalkenes with high yield (up to 90%) in the presence of alkylaluminum dichlorides as a catalyst is shown [37].

Acetylene hydrocarbons, in contrast to alkenes, do not undergo reactions of electrophilic addition of thiols, but they react with thiolates [9, 38]. Unsaturated compounds containing electron-acceptor substituents at the double bond, like esters of acrylic acid [39], maleic anhydride [40] etc. interact with thiolates the similar way.

The interconnection is found between the rate constant ($lg\,k$) of addition of thiols to the double bond of maleic anhydride catalyzed by triethylenediamine, the acidity of thiols (pK_a) and relative nucleophility (n) of conjugated thiolate-anions calculated from $lg\,k$ and pK_a values (Table 1.1) [40]. When decreasing n, the value $lg\,k$ is seen to lower down symbatically. The values n and pK_a change antibatically. The distinction is obviously connected with different structure of butyl group altering pK_a value of isomeric butanthiols and *n*-butanthiolates [10]. The n value for benzylthiol and 2-furfurylthiol is notably higher despite of their lower acidity, evidently because of electron-accepting residues in these molecules.

Table 1.1. Reactivity parameters of different thiols in the reaction with maleic anhydride in the presence of triethylenediamine and *p*-xylene (t = 27°C)

Thiol	pK_a	$lg\,k$	n
n-butylthiol	12.6	2.11	1000
iso-butylthiol	12.9	1.38	380
tert-butylthiol	13.1	1.13	340
benzylthiol	11.8	2.98	1200
2-furfurylthiol	11.3	3.55	1400

Thiols are able to generate not only thiolate-ions but also thiyl radicals, which arise either due to heating [12,41], UV and X-ray irradiation [12, 42] or interaction with free-radical initiators such as peroxides[9, 10, 13]. Radical addition of thiols to alkenes (homolytic thiylation) usually runs through 3 steps with low activation energy ($\boldsymbol{E_a}$):

1. Initiation: $RSH \rightarrow RS^{\bullet}$ (k_1),
2. Attack: $RS^{\bullet}$ + alkene $\rightarrow A^{\bullet}$ (k_2),
3. Chain transfer: $A^{\bullet} + RSH \rightarrow AH + RS^{\bullet}$ etc. ($\boldsymbol{k_3}$).

Chain termination proceeds via radical recombination: $RS^{\bullet} + RS^{\bullet} \rightarrow RSSR$; $A^{\bullet} + A^{\bullet} \rightarrow A_2$; $A^{\bullet} + RS^{\bullet} \rightarrow ASR$. Steps 2 and 3 are shown to be exothermic. This gives rise to rapid chain propagation. Its kinetic length is commonly high (not less than 10^3 turns). The total rate of the process is determined by the chain propagation rate. For example, in the case of interaction of *n*-butanthiol with 1-pentene the following values for the rate constants were found: $k_2 = 7\cdot10^6$ l/(mole·s); $k_3 = 1.4\cdot10^6$ l/(mole·s). Among thiols, the best radical donors are the compounds with the mostly pronounced acidic function [36].

Thiyl radical $RS^{\bullet}$, like other free radicals, always attack the terminal atom C of the double bond of alkene resulting in addition product against the Markovnikov rule. The apparent reason is the less activation energy (E_a) needed for secondary radical formation during attack [12, 36]. So, the result is that $RS^{\bullet}$ attacks terminal C=C bonds approximately 50 times faster than inner bonds.

Free-radical addition of dodecanethiol to phenylene, vinyl ether of *n*-butanthiol, cyclopentene and others [43], as well as the addition of ethanethiol to thiophene in the presence of Fe^{2+} cations, H_2SO_4 and *tert*-butanol [44] is described:

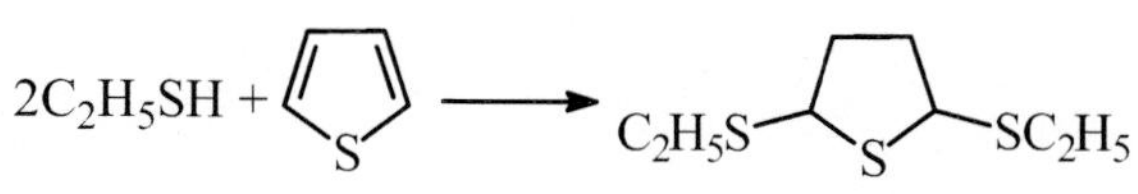

(1.17).

Simultaneous oxidation of thiol and alkene also proceeding via free-radical mechanism [45] gives β-oxyalkylsulfoxides:

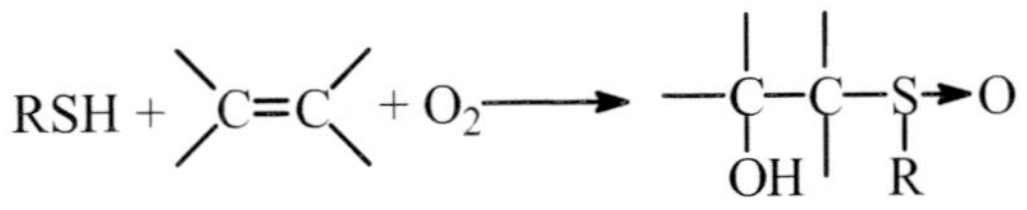
(1.18),

while the oxidative addition of thiol to phenylacetylene results in semi-acetal of phenylglyoxal [46]:

$$C_6H_5C{\equiv}CH \xrightarrow{RS^{\cdot},\ O_2} RS{-}CH(OH){-}C(O){-}C_6H_5 \quad (1.19).$$

Oxidation of thiols with oxygen leads to formation of dialkyldisulfides [43]:

$$2RSH + 0.5O_2 \longrightarrow RSSR + H_2O \quad (1.20),$$

proceeding through the initial transformation of thiol to thiolate-anion, subsequent formation of $RS^{\bullet}$ radical from thiolate-anion via electron transfer to O_2 molecule and the final dimerization of $RS^{\bullet}$ radical to dialkyldisulfide. Accordingly, the rates of these reactions increase in the presence of a base, for instance alcoholate-anion:

$$RSH + R'O^{-} \rightleftharpoons RS^{-} + R'OH \quad (1.21),$$

$$RS^{-} + {\cdot}O{-}O{\cdot} \longrightarrow RS^{\cdot} + {\cdot}O{-}O^{-} \quad (1.22)$$

$$2RS^{\cdot} \longrightarrow RSSR \quad (1.23)$$

Being formed peroxide anion oxidizes thiol:

$$RSH + {\cdot}O{-}O^{-} \longrightarrow RS^{\cdot} + HOO^{-} \quad (1.24),$$

$$RSH + HOO^{-} \longrightarrow RS^{\cdot} + HO^{-} + {\cdot}OH \quad (1.25)$$

$$RSH + {\cdot}OH \longrightarrow RS^{\cdot} + H_2O \quad (1.26)$$

The rate of thiol oxidation to dialkyldisulfides rises when increasing their compatibility with reaction medium [47] that is achieved by different solvents (Table 1.2).

Table 1.2. The rates of *n*-butanol oxidation with oxygen in the presence of sodium methylate depending on the nature of the solvent

Solvent	Oxidation rate constant, $k\cdot10^3$, min^{-1}	Relative oxidation rate
Methanol	3.22	1
Tetrahydrofuran	116	36
Dioxane	289	90
Dimethylacetamide	936	291
Dimethylformamide	1097	334

Thiol oxidation is catalyzed by such metal ions as Fe^{3+}, Cu^{2+} etc. The increase of metal concentration leads to proportional rise of the reaction rate [9]. Deep oxidation of thiol by nitric acid gives alkylsulfo-acids with quantitative yield [15].

It is worth noting that thiols also can actively transfer the oxidation chain [9, 10, 12, 13]. For this reason they are applied widely as accelerants of polymerization of unsaturated compounds, antioxidants, radiation protectors etc. (see Section 1.3).

1.3. Fields of Application of Thiols and Their Derivatives

The following chemicals are produced basing on individual thiols and, in rare instances, blends of thiols.

- Perchloromethylthiol, $ClSCCl_3$ is an active insecticide, fungicide and herbicide [7]. Its force of influence is equal to potassium cyanide, but its application is restricted due to high toxicity towards warm-blooded animals. Perchloromethylthiol underlies the subsequent synthesis of polychlorinated sulfides and sulfamides, being used as the preparations like captan, phtalan, euparen, captofol and others. These chemicals are less toxic and are effective agents against apple scub, grey mould on strawberry and stone crops, grain crops deseases etc. [48]. Similarly, perchloromethylthiol and cyanide salts are used to synthesize methylisocyanates that suppress weeds [7].
- Various S-derivatives of *sym*-triazines (products of interaction of chlorotriazines with individual sodium thiolates) represent the wide range of low-toxic herbicide preparations symazine, atrazine, prometrin, desmetrin, metributin and others aimed to fight with weeds of rice, cabbage, carrot, grain crops, potato and pineapple [48]. N-containing derivatives of methylmercapto-*sym*-triazine possess multiform antiviral activity [49]. C_1-mercaptoaminotriazines and their derivatives are able to promote vulcanization of synthetic rubbers enhancing operating performance of the resulted products [50].
- C_1-C_3-thiochlorocarbonates, RSC(O)Cl have the synthetic ways mentioned above. They can damage the respiratory system of harmful insects much more effectively than initial lower thiol homologues, which also show the similar acaricide activity.
- C_1-C_3-thioethers of methylenglycoles, $R'OCH_2OC_2H_4SR$ exhibit insecticide activity [51].

- C_4-C_6-alkylvinylsulfones, $RSO_2CH{=}CH_2$ are oxidized products of interaction of sodium thiolates with acetylene. They possess herbicide and acaricide activity [38]. Alkylvinylsulfides, $CH_2{=}CHSR$, were synthesized earlier [52]. Basing on them alkylvinylsulfones and alkylvinylsulfoxides, $CH_2{=}CHS(R)\rightarrow O$, were synthesized. Alkylvinylsulfoxides have a good repellent effect on a number of insect species; they are also applied for curing the mange of cattle as well as the helminthiasis of animals and birds [15]. On treating alkylvinylsulfides with esters of dithiophosphoric acid, corresponding derivatives are produced, $(C_2H_5O)_2P(S)SC_2H_4SR$, possessing insecticide properties [52]. The chemical $(C_2H_5O)_2P(S)OC_2H_4SC_2H_5$ of this type is the base for well-known insecticide preparation mercaptophos being the reference remedy against cotton pests [17].
- Various biologically active substances including natural ones, that contain –SH groups. Many of them are involved in synthesis and decomposition of fats and proteins in living organisms, for example cystein (γ-mercaptoamino acid), cystamine (β-mercaptoethylamine) is a component of coenzyme A, methionine (methyl ester of homocystein) and others [53]. Cysteinproteinases (enzymes) provide the formation of disulfide bridges fixing the conformation of peptide and protein macromolecules. Cystein residue in glutathione tripeptide is account for its important role in the biochemistry of redox processes involving free radicals [54]. Methionine is a highly effective protein additive for fodder. For that reason the industrial method of its production was developed via treatment of akryl aldehyde by methanethiol in the presence of acetylperoxide with subsequent conversion of methylmercaptopropionic aldehyde into 4-hydantion and hydrolysis of the latter [5]. Symmetric dialkylsulfides obtained from C_4-C_6-thiols blends revealed themselves as stimulators accelerating the growth of animal's hair for 2-3 times [55].
- *Bis*-(ethylsulfonyl)-dimethylmethane, $(CH_3)_2C(SO_2C_2H_5)_2$ being the product of oxidation of dimethyldi(ethylmercaptal) with potassium permanganate, is a popular somnifacient sulfonal [1]. The considerable effect of C_3-C_6-*iso*-alkylthiols on the central nervous activity of animals [56].
- C_2 and C_3-mercaptoderivatives of 1,3-dioxolans show significant antibacterial activity against staphylococcus and streptococcus infections as well as against blue pas and colon bacillus [57]. Such an activity is peculiar to 1,3-diheterocycloalkanes, which are the interaction products of, for example, C_2-C_3-thiols with 4-halogenomethyl-1,3-dioxolans [58]. In general, a large number of medicines for cardiovascular diseases is synthesized on the base of thiols [59, 60].
- Alkyl-*bis*-tetramethylenesulfides, $(C_5H_{13} - C_{12}H_{25})S(CH_2)_4S(C_5H_{13} - C_{12}H_{25})$ with the concentration about $2\cdot10^{-2}$ mole/l noticeably constrain polymethylmetacrylate chains breaking during their thermal-oxidative destruction [61]. At the higher concentration *bis*-sulfides begin transferring the oxidation chains and polymethylmetacrylate destruction progresses, in opposite. Polymethacrylate is heat-stabilized by introduction of 0.1 mass % of *iso*-dodecanthiol. The same addition to either polyethylene or polypropelene has little effect [8]. However when combined with soot the considerable synergetic effect is observed.

- Monothioethyleneglycol and monothiopropeleneglycol contained in ethylpropelene rubber or low-density polyethylene deactivate metal-organic catalysts and neutralize their destructive effect [8].
- Didecylsulfide, $(C_{10}H_{21})_2S$ is produced by oxidation of corresponding thiol. Along with other dialkylsulfides, it is rather weak antioxidant of polyethylene, but sharply intensifies its strength when combined with soot, diphenoquinone or polyacene [8].
- Sn, Sb and Cd salts of C_5-C_{22}-thiols are able to absorb HCl and to inhibit radical oxidation processes. They are effective stabilizers of polyvinylchloride [8].

At the same time, solutions of different dialkylsulfides in organic solvents can recover Ag, Hg, Pt, Au and Pd cations from acid extracts of mineral ores containing corresponding metals [15]. The nature of the solvent strongly affects the efficiency of metal recovery. For example, Ag^+ extraction coefficient increases from 12 to 17300 when changing *n*-heptane to CCl_4, benzene or $CHCl_3$. Efficiency of sulfide extracting agents increases when oxygen atoms are introduced into their molecules [62] by means of ketosulfides synthesis through the reaction of thiomethylation of ketones described above [31]. Extraction of silver [63], gold [64], palladium and platinum [65] is investigated using some ketosulfides.

Dialkylsulfides and dialkyldisulfides are the active flotation reagents for sulfide copper-nickel and copper-molybdenum ores [66]. The most effective is asymmetrical *iso*-C_6-phenylsulfide [67], which provides 80% yield of the concentrate with consumption equal to 100 g per ton of crushed ore. Such a yield corresponds to the level of extraction of copper, nickel and molybdenum sulfide equal to 75-95%. At the same time symmetrical dialkylsulfides, for instance di-*n*-octylsulfide, suppress flotation of the other sulfide minerals: pyrite, chalcopyrite and sphalerite [68]. Effective flotation properties of ketosulfides are also reported [69].

Substituted sulfides containing polysulfide bridges, in particular di-*tert*-C_4 and di-*tert*-adamantylpolysulfides can serve the effective additives for lubricating oils diminishing friction and abrasion [70].

- Aminothiols (for example $H_2NC_2H_4SH$), aminosulfides, aminodisulfides and related derivatives of 2,6-di-*tert*-butyl-4-methylphenol (ionol) can effectively prevent oxidation of hydrocarbon fuels (gasoline, diesel oil, jet engine fuel) and formation of sediments in them [17,71].
- Thiomethylenealkylphenols contain thiomethylene groups in *ortho* and *para* positions of aromatic ring. They increase antioxidative and anticorrosive properties of lubricating oils [72].
- Alkyl esters of 2,5-dialkylbenzylthioacetic acids lower steel corrosion from 360 to 8-12 g/m^2 when added to lubricating oil M-11 in the amount of 1-2 mass % [72].
- Phosphoroorganic thioethers of alkylsulfides, $RSSCH_2P(O)(OR')_2$ increase antioxidative and anticorrosive properties of lubricating oils [72].
- Bis-(alkylamide)sulfides, $(R^1R^2NC(O)CH)_2S$ [73], amides of thiocarboxylic acids [74, 75] are antioxidation and antideterioration additives to oils.
- Alkylmethyldithiobenzoates, $C_6H_5C(S)SCH_2NR^1R^2$ are antioxidation and antideterioration additives to lubricating oils [76].

- Alkylthioethylaminophenylalkylthioketones, $C_6H_5CH_2C(S)NH(CH_2)_2SR$ and $C_6H_5C(S)NH(CH_2)_2SR$ are additives increasing anticorrosive and lubricating action of engine oils [77].
- Heterocyclic thiols synthesized by a specific way through the interaction of aliphatic and aromatic amines with CS2 [8, 50], namely mercaptobenzothiazol and its derivatives formed by oxidation (2,2'-dithio-*bis*-thiazol) and by aminomethylation (diethylaminomethylmercaptobenzothiazol [78]), 2-mercaptobenzimidazol, 2-mercaptoimidazol etc are highly effective deactivators of hydroperoxides (products of thermal oxidation of many polymer materials) and the promoters of vulcanization of various rubbers.

Finally, the wide possibilities of thiols application as synthons in fine chemical synthesis are described [79]. Due to the strong specific odor noted in the beginning of this review C_2-C_5-alkylthiols find an application as an odorant of industrial and domestic fuel gas in Russia and abroad [5, 80].

This section testifies the wide application of thiols in various areas of industry, agriculture, medicine etc. As a result, raw materials sources and methods of thiol production are of great importance.

1.4. Production of Thiols by Synthetic Methods and from Natural Raw Material Sources

The general sources of alkylthiols are their synthesis based on available organic compounds and their isolation from sulfurous oils and gas condensates [15, 17].

Methods of alkylthiols synthesis include the following.

- Interaction of alkyl halides with the excess of sodium (or potassium) hydrosulfide at the temperature 60-160°C and at slightly elevated pressure (2-3 MPa) [5, 81]:

$$RCl + NaSH \rightleftharpoons RSH + NaCl \qquad (1.27).$$

- Addition of H_2S to 1-alkenes via free-radical mechanism [12, 82]:

$$RCH{=}CH_2 + H_2S \xrightarrow{h\nu} RCH{=}CH_2 + (H^{\cdot} + {:}SH) \longrightarrow$$
$$\longrightarrow RC^{\cdot}HCH_2SH \xrightarrow{H_2S} RCH_2CH_2SH + {^{\cdot}}SH \qquad (1.28).$$

- Addition of H_2S to 1-alkenes via carbonium ionic mechanism [5, 12]:

$$RCH{=}CH_2 + H_2S \xrightarrow{H^+} RC^+H{-}CH_3 + H_2S \longrightarrow RCH(SH){-}CH_3 \qquad (1.29).$$

- Reaction of alcohols with H_2S in the presence of catalysts (Ct) of acid (aluminum oxide, aluminum silicates and mineral acids supported on them) or base (hydroxides, carbonates and tungstates of K or Na) type [83, 84]:

$$ROH + H_2S \xrightarrow{Ct} RSH + H_2O \qquad (1.30).$$

One of the variants of this reaction is the interaction of corresponding aldehydes, ketones or carboxylic acids with H_2S in the presence of hydrogen and catalytic amount of Ni, Co or Mo sulfide [83].

- Reaction between alkyl halides and sodium thiosulfate with subsequent decomposition of Bunte salt with sulfuric acid [17]:

$$RCl + Na_2S_2O_3 \xrightarrow[-\ NaCl]{} RSSO_2ONa \xrightarrow[-\ Na_2SO_4,\ SO_2\uparrow,\ S\downarrow]{+\ H_2SO_4} RSH \qquad (1.31).$$

- Interaction of alkyl halides with thiourea and subsequent decomposition of S-substituted *iso*-thiouronium salt with sodium hydroxide [17]:

$$RCl + S{=}C(NH_2)_2 \longrightarrow RS^+{=}C(NH_2)_2Cl^- \longrightarrow$$
$$\xrightarrow[-\ H_2O]{+\ NaOH} RSH + (NH_2CN)_n \qquad (1.32).$$

- Interaction of *tert*-alkanol with thiourea and concentrated hydrobromic acid and subsequent decomposition of the intermediate product with sodium hydroxide [17]:

$$ROH + S{=}C(NH_2)_2 + HBr \longrightarrow H_2N^+C(NH_2)_2SRBr^- \longrightarrow$$
$$\xrightarrow{+\ NaOH} RSH \qquad (1.33).$$

Among methods of alkylthiols synthesis listed above the following gained the industrial significance [5, 81]: exchange reaction of alkyl halides with sodium hydrosulfite (for synthesis of *n*-alkylthiols), addition of H_2S to alkenes in the presence of acid catalyst, preferably aluminum chloride or C_1-C_4-alkylsulfo acids (for synthesis of *tert*-alkylthiols), and to the smaller extent, interaction of alcohols with H_2S in the presence of A12O3 (for synthesis of alkylthiols of various structure). In the last case the most expensive raw materials (alcohols, especially higher) and harsh synthesis conditions (t up to 300°C, P = 4-5 MPa) are used achieving comparably low yield of target thiols (50-70%) and noticeable amount of by-products (dialkylsulfides, dialkyl ethers etc. [84]).

With the evolution of the technology of catalytic processes of alkylthiols production their differentiation on product types has occurred. To synthesize C_6-C_{15}-*tert*-alkylthiols the low temperature (0-10°C) is the most suitable for the reaction of corresponding isoalkenes with H_2S in the presence of $AlCl_3$ at the air pressure [85]. Keeping these conditions allows achieving 80-85% yield of target thiols in the reaction mixture. To obtain C_2-C_5-alkylthiols, the sulfides of Ni, W or V as well as phosphorotungstic or other heteropoly acids supported on silica gel should be applied as catalysts. The process should be conducted at t = 300-350°C, P = 1.0-1.4 MPa and the conversion of initial alkenes equal to 85-100% [83].

Despite the relatively high cost and the ecological danger of production of alkylthiols by synthetic methods, the year volume of their production in the world is about 200,000 tons [6]. Only the Pennvalt company (USA) has the capacity for alkylthiols production up to 100,000 tons per year. This company tells that the world market's needs will stably increase for 6-8% per year.

Alkylthiols are the most important group of sulfur-containing organic chemicals produced at the industrial level [83]. Methylthiol (up to 50,000 t/year) is used in the production of protein feed additive (methionine). C_2-C_5-thiols are used as an odorant for fuel gas and as the raw material for the production of plant's chemical protectants and domestic chemicals. C_6-C_{15}-thiols are the promoters for rubber vulcanization, the raw material for the synthesis of antioxidants of polymeric materials. C_{10}-C_{18}-thiols are the inters in surfactants' Foreign companies manufacture not less than 12 types of synthetic alkylthiols C_1-C_{18} [6].

Another large-tonnage source of alkylthiols is sulfurous oils constituting the prevailing part of ones produced in the world (especially in Russia, in Ural-Volga region, Caspian Basin, Western Siberia and others), and sulfurous gas condensates. Total amount of sulfur contained in them varies in a very wide range from 0.5 to 10 mass %. This sulfur is generally contained in the molecules of sulfur-organic compounds such as mercaptans, disulfides, sulfides, thiophenes and compounds of unidentified structure (so-called residual sulfur), as well as in the elemental sulfur dissolved in oil or gas condensate [86].

Investigations of structural-group composition of oil's distillate fractions (boiling in the range 40-450°C) carried out in the Institute of Organic Chemistry, Ufa Research Centre, Russian Academy of Science by Prof. N.K. Lyapina's group allowed to develop their chemical classification (by sulfur compounds) [15]. Mercaptan oils contain thiols, sulfides and thiophenes in comparable amount. Thiophene oils contain benzo and dibenzothiophenes, sulfide amount is ≤ 10%. The content of condensed molecules, S-organic substances and hydrocarbons is shown to increase when increasing the boiling point of distillates [87]. Rise of molecular mass of these compounds is determined by both increase of cyclization extent and lengthening of alkyl chain.

The high content of mercaptan sulfur is characteristic for oils and gas condensates related to carbonate deposits [88] including oils of Samara region (0.45 mass %), Markovskoye (~ 0.6 mass %) and Ishimbayevskoye (0.3 mass %) oil fields, as well as oils and gas condensates from Caspian Basin (0.14-0.60 mass %) [89-92].

Structural-group composition of mercaptans containing in the hydrocarbon raw material mentioned was studied using different reaction ability of primary, secondary and tertiary alkylthiols in various reactions [93-95] that provided some error in the data obtained [86]. Nevertheless, resulted from these and earlier studies, the conclusion is done about the content of all possible C_1-C_9-alkylthiols in gasoline fractions of oils and condensates. The most of those alkylthiols is represented by isomers with alkyl chain branched at α-position relative to S atom [15].

The accuracy of analysis increased due to development and application of negative-ion mass spectrometry with resonant electron capture [96, 97] which allowed determining the individual and structural-group composition of thiols and disulfides without their isolation from mixture with hydrocarbons, particularly in various gas condensates from Caspian Basin [98, 99]. Thus, contained in the distillate with b.p. 180-350°C of Astrakhan gas condensate, thiols was discovered (0.524 mass %) with substituents of cyclic (60%) and regular (40%) structure composed of 6-9 C atoms. The structural-group composition of different fractions of Orenburg, Karachaganak and Astrakhan gas condensates is listed in [90], proving the considerable amount of C_1-C_3-alkylthiols contained in them (Table 1.3).

Table 1.3. Individual composition of mercaptans from Orenburg, Karachaganak and Astrakhan gas condensates (rel. %)

Compound	Orenburg* fract. 40-192°C	Karachaganak* fract. 65-520°C	Astrakhan* fract. 71-560°C
CH_3SH	0.2	0.5	13.4
C_2H_5SH	12.7	16.8	17.2
C_3H_7SH (2 isomers)	52.6	44.7	20.6
C_4H_9SH (4 isomers)	27.6	13.8	20.7
$C_5H_{11}SH$ (8 isomers)	5.6	4.2	10.5
$C_6H_{13}SH$ (4 isomers)	1.0	0.9	5.2
$C_7H_{15}SH$ (3 isomers)	0.1	2.2	2.0
n-$C_8H_{17}SH$	traces	5.3	1.5
n-$C_9H_{19}SH$	traces	2.6	1.2
n-$C_{10}H_{21}SH$	-	0.6	0.3
n-$C_{11}H_{23}SH$	-	0.3	traces
n-$C_{12}H_{25}SH$	-	traces	traces
n-$C_{13}H_{27}SH$	-	traces	-
n-$C_{14}H_{29}SH$	-	traces	-
n-$C_{15}H_{31}SH$	-	traces	-
cyclo-C_5H_9SH	traces	0.2	0.6
$C_6H_{11}SH$**	traces	}	1.5
$C_7H_{13}SH$***	traces	4.2 }	2.2
$C_8H_{15}SH$****	-	}	} 1.8
$C_9H_{17}SH$	-	3.0 }	} 0.8
$C_{10}H_{19}SH$	-	0.7 }	} 0.3
C_6H_5SH	-	traces	0.2
Total	100.0	100.0	100.0

Notes: *Content of total and mercaptan sulfur, mass %: Orenburg 1.7 and 0.63, Karachaganak 0.62 and 0.18, Astrakhan 0.96 and 0.15;

**Cyclohexylthiol, 2-methyl and 3-methylcyclopentylthiols;

*** Cycloheptylthiol and 2-methylcyclohexylthiol;

****2-Ethyl and 2,5-dimethylcyclohexylthiol.

Resulted from studying the individual composition of thiols containing in the gas condensates of Caspian Basin with the use of series of spectral methods (negative-ion mass spectrometry with resonant electron capture, IR spectroscopy, ^{1}HNMR and ^{13}CNMR spectroscopy) the following general regularities was established [90, 99].

The study of raw hydrocarbon material produced at the fields of Caspian Basin (astrakhanskoe, orenburgskoe, karachaganakskoe, tengizskoe, zhanazholskoe) revealed the existence of a wide spectrum of sulfur-containing compounds along with aliphatic and aromatic hydrocarbons [15, 86, 87, 89-92, 98-105]. So, an amount of mercaptan sulfur in gas condensates of the fields listed is 26.7-57.6% from its total content including 1% from Orenburg gas condensate [106]. Sour natural gas purified at Orenburg gas refinery contains 400-500 mg/m^3 of mercaptan sulfur [107].

1. 38 alkylthiols were identified in gas condensates studied, 30 of them are aliphatic with the number of carbon atoms from 1 to 14, 8 are cyclic being the derivatives of cyclopentene and cyclohexane with the number of carbon atoms from 5 to 11, and finally cycloheptylthiol.
2. Gas condensates contain almost all possible light thiol. Compounds of regular structure prevail among the heaver thiols. Among aliphatic thiols the most probable are the isomers branched at the position 2 with respect to sulfur atom.
3. For *iso*-structured aliphatic thiols the number of branching does not exceed 1-2 C atoms.

The presence of sulfur-containing compounds, especially mercaptans, being much less thermally stable than hydrocarbons [15, 17], complicates processing of raw hydrocarbon material and negatively affects the properties of the products synthesized from this raw material [81, 108]. To isolate alkylthiols from liquid and gaseous raw hydrocarbon material of natural origin absorption, adsorption and chemical technologies are applied [109-112], although this sorting is arbitrary in many cases.

The most popular and effective *a b s o r b e n t s* of alkylthiols are aqueous solutions of KOH or NaOH due to acidic properties of alkylthiols as mentioned in Section 1.1. The constant of distribution of thiols (K_q) between these solutions and hydrocarbon phase [107] depends (according to equation (1.34)) on relative solubility of thiol in these phases (K_p), its ionization constant (K_{RSH}) and concentration of KOH (NaOH) ($[OH^-]$) and H_2O ($[H_2O]$) in the solution.

$$\boldsymbol{K}_q = \boldsymbol{K}_p \cdot \frac{\boldsymbol{K}_{RSH}[\mathrm{OH}^-]}{\boldsymbol{K}_{H_2O}[\mathrm{H_2O}]} + \boldsymbol{K}_p \qquad (1.34),$$

where

$$\boldsymbol{K}_p = \frac{[\mathrm{RSH}]_{\mathrm{H_2O}}}{[\mathrm{RSH}]_{\mathrm{HC}}} \qquad (1.35),$$

$$\boldsymbol{K}_{RSH} = \frac{[\mathrm{H}^+][\mathrm{RS}^-]}{[\mathrm{RSH}]} \qquad (1.36),$$

$$\boldsymbol{K}_{H_2O} = \frac{[\mathrm{H}^+][\mathrm{OH}^-]}{[\mathrm{H_2O}]} \qquad (1.37).$$

Values of K_p calculated from equation (1.34) for some RSH at t = 20°C are listed in Table 1.4. It is evident that K_p value decreases with raising the molecular mass of RSH and the concentration of KOH (NaOH) solutions. K_p decreases also with temperature increase [107]. The stated influence of the concentration of KOH (NaOH) solutions at K_p is obviously caused not only by relative decrease of H_2O content but also by a known [113] effect of salting-out thiols with K (Na) thiolates (in this case), which possess larger electric charge and are hydrated more actively. Salting-out constant marked as m in equation (1.38) increases when passing from CH_3SH to C_4H_9SH and from KOH to NaOH solution (Table 1.5).

$$\lg \frac{K_p}{K_p^{H_2O}} = -m \cdot [\text{KOH(NaOH)}] \qquad (1.38).$$

The interconnection is found [107] (by means of equations 1.39-1.41) between the boiling point (T_b, K) of aliphatic regular thiols and m value in solutions:

for KOH, $\lg \boldsymbol{m} = -5.31 + 0.011\ \boldsymbol{T_b}$ (1.39),

for NaOH, $\lg \boldsymbol{m} = -3.37 + 0.007\ \boldsymbol{T_b}$ (1.40),

as well as the solubility in water,

$$\boldsymbol{K}_p^{H_2O} = -12.36 + 0.02115\,\boldsymbol{T}_b + \frac{4695}{\boldsymbol{T}} - 8.88\frac{\boldsymbol{T}_b}{\boldsymbol{T}} \qquad (1.41),$$

where T is the temperature of thiol absorption.

Table 1.4. K_p values of thiols in aqueous KOH and NaOH solutions (t = 20°C)

Concentration of hydroxide solution, mole/l	C_2H_5SH		n-C_3H_7SH		n-C_4H_9SH	
	KOH	NaOH	KOH	NaOH	KOH	NaOH
0.5	6.34	6.06	-	3.28	-	1.88
1.3	6.16	5.08	-	-	-	-
2.0	-	5.02	3.26	-	1.58	0.97
2.8	5.93	4.23	3.06	-	1.39	-
3.0	-	3.38	-	1.94	-	0.64
4.0	-	3.32	-	-	-	0.42

Joint solution of equations (1.34-1.41) allowed calculating the material balance of the process of thiols' absorption by KOH or NaOH water solutions with their concentration lying in the range of 0.5-6.0 mole/l and t = 0-50°C [107] with the accuracy needed for technical project. According to the data on solubility of thiols of regular structure listed above their absorption with KOH solution is more preferable being more effective than with NaOH solution (about 1.3 times in the same conditions). Moreover, potassium thiolates are better soluble in water in comparison with similar Na salts that allows using KOH solutions with higher concentration.

Table 1.5. Values of distribution constants in water ($K_p^{H_2O}$), dissociation constants (K_{RSH}) and salting-out constants (m) for some thiols in aqueous solutions containing approximately 3.5 mole/l KOH (NaOH)

Thiol	$K_p^{H_2O}$	K_{RSH} ($\times 10^{11}$)	m ($\times 10^2$)	
			KOH	NaOH
CH_3SH	12,85	4,20	0,6	3,8
C_2H_5SH	5,68	2,52	1,3	6,6
н.C_3H_7SH	3,82	2,26	2,5	9,2
н.C_4H_9SH	2,24	2,21	6,0	18,5

However, KOH is much more expensive than NaOH. As a result, exactly the latter is used at Russian gas refineries for isolation of C_2-C_5-thiols from light fractions of sulfurous gas condensate [6, 110]. The interference of thiols of regular and iso structures during absorption of their mixtures is studied. The equations are revised that describe the material balance of this process allowing to calculate the extraction level of thiols' mixtures of various composition from liquid, liquefied and gaseous hydrocarbons. On raising the temperature the value of K_q as well as K_p noticeably decreases due to more intensive hydrolysis of Na thiolates that is used for isolation of thiols from alkali solution [6, 107]. Investigation of hydrolysis kinetics of thiolates mentioned above resulted in equation (1.42) obeying Arrhenius law in the range t = 70-120°C and linking the constant of this reaction (K_H) with NaOH concentration [NaOH] in the solution. The mathematical model is developed for calculation of the material balance of the process.

$$K_H = A / [\mathrm{NaOH}]^n \quad (1.42).$$

According to equation (1.42) dependence of K_H on [NaOH] is expressed in logarithmic scale by the straight line with the slope (n) equal to 1.5 and the A value listed in Table 1.6. The observed increase of A value (1-2 orders of magnitude) for homologues and isomers of Na thiolates indicates the big difference between their hydrolysis ability. To raise thiolate mixture's hydrolysis rate and, correspondingly, the reactor productivity high temperature and minimal [NaOH] are necessary.

Table 1.6. Preexponential factor *A* for sodium salts of C_1-C_4-thiols in the equation (1.42)

Thiol	A
CH_3SH	3.50
C_2H_5SH	9.30
n-C_3H_7SH	30.0
iso-C_3H_7SH	47.6
n-C_4H_9SH	57.5
iso-C_4H_9SH	148.7
sec-C_4H_9SH	156.6
tert-C_4H_9SH	332.5

Dilution of thiolates' solution with water is established [6, 107] to increase the rate of heir hydrolysis and provide almost 95% level of extraction of C_2-C_4-thiols from alkali solution. Moreover, simultaneous hydrolysis of thiolates containing in products of demercaptanization of various hydrocarbon flows at Orenburg gas refinery (Table 1.7) allows achieving high level (96% and higher) of desorption of the thiols (except Na methyltiolate). The best result show the flows I + IV + V + VI combined with II + III with volume ratio more than 1 : 4.

Table 1.7. Composition of mixtures of sodium thiolates produced during processing of liquid-gas hydrocarbon flows at Orenburg gas refinery

Sodium thiolates	Content of sodium thiolates in raw material flows, rel. %					
	Gas	Stable condensate	Condensate fraction, b.p. ≥ 130°C	Condensate of thiols' drying vapors	Long fraction of light hydrocarbons	Propane-butane fraction
	I	II	III	IV	V	VI
C_1	34-24	0.3-3.0	3-5	10-20	40-60	80-85
C_2	20-25	25-35	35-45	85-75	50-40	15-10
iso-C_3	35-30	40-35	45-30	3-5	15-10	1-5
n-C_3	5-15	15-5	5-10	2-1	5-2	–
sec + *iso*-C_4	15-10	2.5-1	15-10	traces	–	–
tert-C_4	traces	1.5-2.5	3-5	–	–	–
n-C_4	traces	5-2	3-1	–	–	–
higher	–	–	–	–	–	–

Based on the regularities stated above the technological flowsheet (Figure 1.1) for absorption of C_1-C_6-thiols from raw hydrocarbon flows with 10% NaOH solution and their isolation after hydrolysis of thiolates' alkali solution is developed and realized [114-116]. In the course of the technological developmental work on the plant production flowsheet it turned out expedient to subject propane-butane fraction of natural gas and hydrocarbon condensate to sweetening only but not the total flow of natural gas [114, 115].

In particular light fraction of gas condensate is preliminary washed with aqueous solution of diethanolamine to extract generally H_2S and CO_2 whereupon it is fed to C-1 extraction still (Figure 1.1). Approx. 10% NaOH aqueous solution cooled to 30-40°C is fed to the top of the extractor by P-1 pump. Thiols are absorbed (extracted) from hydrocarbon raw material in the column C-1 yielding Na thiolates. Raw material purified from thiols comes to S-1 separator and, then, to the stock tanks.

Alkali solution saturated with thiolates is removed from C-1 still, warmed in E-1 heat exchanger up to 80-90°C and fed to the top of C-2 regenerator. Before E-2 heat exchanger the alkali solution is diluted for 20 vol. % by circulated water from separator S-2. In the still C-2 alkali solution is warmed by steam up to 110-125°C. As a result, Na C_2-C_4-thiolates are mainly (according to the Table 1.6) hydrolyzed into thiols, which are blown out from the solution.

Desorbed thiols with water vapor come to condenser E-3 and then to separator S-2. Water containing dissolved thiols (up to 0.5 g/l) is removed from S-2 and fed to E-2 inlet. Thiols from the top of S-2 are forwarded to the stock tanks. Regenerated alkali solution is returned to C-1. A portion (10-20 vol. %) of circulating alkali solution comes into reactor C-3 for drastic oxidative regeneration (at the heterogeneous catalyst UVKO-2 that is fibrous carbon material containing 0.02-0.05 mass. % of Co phthalocyanine being the oxidation catalyst) resulting in transformation of thiolates remained in the solution, generally Na methylthiolate into dialkyldisulfides, and transformation of Na sulfide into Na thiosulfate [116]. Outgoing air containing traces of C_1-C_2-thiols from the top of C-3 after passing the separator S-3 and the flame arrestor is forwarded to incineration.

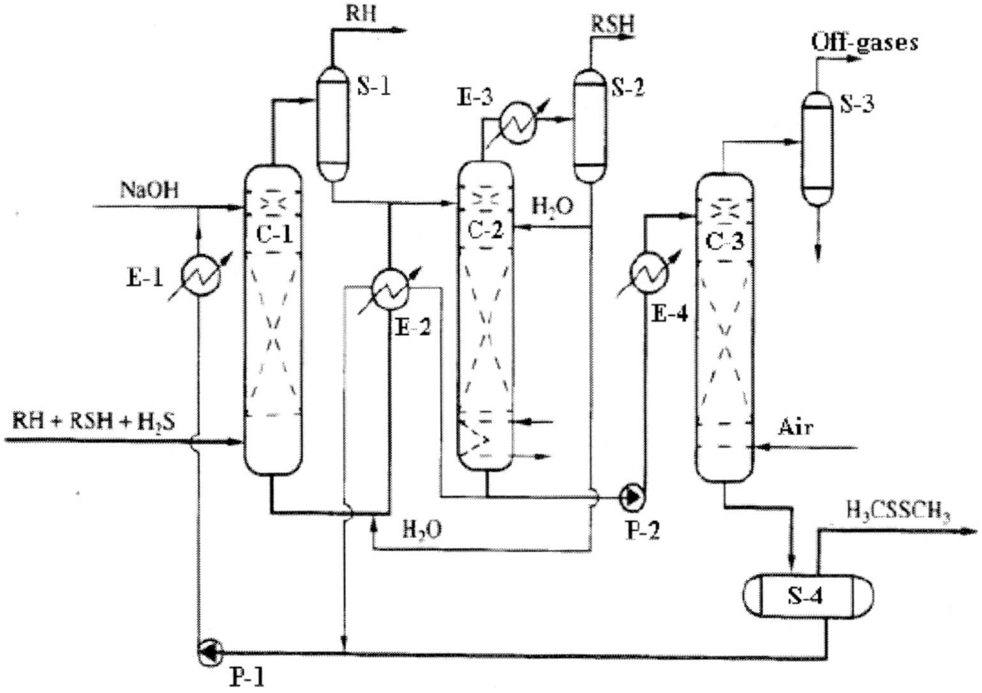

Figure 1.1. Schematic flowsheet of thiols' mixture extraction from raw hydrocarbon material. C-1 – extractor (absorber); C-2 – regenerator; C-3 – reactor; P-1,2 –pumps; S-1,2,3,4 – separators; E-1, 2, 3, 4 – heat exchangers.

Dialkyldisulfides produced during the oxidation of indicated portion of alkali solution contain up to 90% of dimethyldisulfide and are much less toxic in comparison with the initial methylthiol. They can be used as inhibitors of coking during the pyrolysis of low-sulfur raw material, and also as a reagent for preliminary sulfurization of hydrofining catalysts (instead of H_2S) [116].

On accumulating Na thiosulfate a portion of alkali solution is periodically removed from the unit and injected into lost circulation horizon [6]. It has been recently found out [111] that regenerated alkali solution can be additionally purified from unhydrolyzed Na thiolates and Na sulfide via air oxidation at 30-40°C with the exposure to ultrasonic waves with the frequency 22 kHz and the power 58 W/cm^2. This exposure provides ultrafine distribution of an oxidant in the solution, enhancement of the surface of interacted phases along with considerable diminishing of the energy consumption in this process.

Operating experience for the unit described above (Figure 1.1) showed that the level of extraction of thiols mixture is stable at the 60% level from their potential content in the raw material. The desorption level is higher than 90% while the steam consumption is 80-100 kg per m^3 of alkali solution [6]. Extracted thiols contain average amounts (in mass. %) of CH_3SH – 0.30, C_2H_5SH – 34.0, *iso*-C_3H_7SH – 43.0, *n*-C_3H_7SH – 9.70, *tert*-C_4H_9SH – 0.80, *iso* and *sec*-C_4H_9SH – 10.70, *n*-C_4H_9SH + higher – 1.50, i.e. 80% of them are ethyl and propylthiols. The mixture of this (and close to this) composition has the characteristics complying with the requirements for odorants of domestic and industrial fuel gas [80]. Application of thiols as odorants will be discussed in details below (Section 1.5).

Treatment of thiol-containing hydrocarbon flows with aqueous solutions of bases weaker than KOH (or NaOH), namely ethanolamines, in particular DEA, does not result in noticeable thiols' absorption especially in the presence of H_2S and CO_2 [107]. Thus, on the absorption unit at Orenburg gas refinery after purification of sulfurous gas with 25% aqueous solution of DEA thiols' averaged content lowers from 450 down to 400-420 mg/m^3 (calculated for sulfur), i.e. for 7-12.5%. At the same time, many organic substances soluble in water are able to physically absorb thiols, for example propylenecarbonate (PC), N-methylpyrrolidone (NMP), tributylphosphate, dimethyl ether of tetraethyleneglycol (DME) and similar alkyl ethers of glycols as well as the ethanolamine aqueous solutions containing them [107, 115]. Constants and heats of dissolution of ethylthiol, methane and *n*-butane in the absorbents mentioned are listed in the Table 1.8.

According to Table 1.8 ethylthiol is best absorbed by DME (K has the lowest value while ΔH has the highest one), and a little worse by TBP and NMP. The same substances can absorb n-butane and its higher homologues, though in considerably less extent than C_2H_5SH that proves the affinity between the latter and thiols. Higher hydrocarbons are actually used as thiol absorbents, but the process proceeds only at low temperatures (below -30°C) when the value of K is lower (Table 1.8), not providing high level of extraction even in this case [110, 115]. This feature, being common for all physical absorbents, and the presence of noticeable amount of hydrocarbons in the desorption gases along with thiols restrict application of the organic substances listed above as the absorbents of sulfurous admixtures.

Table 1.8. Constants (*K*, kPa/mole fraction) and heats (Δ*H*, kJ/mole) of dissolution of ethylthiol, methane and *n*-butane in some organic substances and methyldiethanolamine (MDEA) aqueous solutions containing them at 20°C

Absorbents	Ethylthiol		Methane		*n*-Butane	
	K	Δ*H*	*K*	Δ*H*	*K*	Δ*H*
Propelenecarbonate (PC)	11.5	28.4	–	–	45.0	14.5
N-methylpyrrolidone (NMP)	5.3	30.6	–	–	23.2	14.8
Tributylphosphate (TBP)	2.5	32.7	–	–	9.9	14.9
Dimethyl ether of tetraethyleneglycol (DME)	1.8	35.0	330	6.2	7.2	15.0
MDEA aqueous solution (50 mass %)	50	24.5	16000	5.7	350	14.2
Aqueous solution A (mol %): MDEA –35, DME – 35	24.0	33.7	5800	5.8	110	14.5
Aqueous solution B (mol %): MDEA –10, DME – 45	6.2	34.0	1600	6.0	40.0	14.9

At the same time, introduction of these substances into ethanolamine solutions, for instance in MDEA solution that can, in contrast to DEA [111, 114], selectively isolate H_2S from the gas, considerably raises its thiol absorption ability (Table 1.8). The industrial tests of such an aqueous organic absorbent were carried out at Orenburg gas refinery. The tests showed the possibility of removing up to 50% of thiols producing gas purified to residual content of H_2S less than 20 mg/m^3 and the desorption gas including 60-65 vol. % of H_2S, no more than 6 vol. % of hydrocarbons and matching the requirements for raw material used for effective processing into elemental sulfur in the Claus process.

Being developed and commercially tested is the alternative technology [110] of thiols isolation from sulfurous natural gas with polysulfide derivatives of DEA, which are formed on introduction (dissolution) of elemental sulfur (up to 1 mass. %) into the 25% aqueous solution of DEA saturated with hydrogen sulfide. During the process up to 80% of thiols contained in the original gas react with polysulfide mentioned giving polysulfide oil which is removed from the process when the absorbent is subjected to regeneration.

Development of the *adsorbtion* method for isolation of thiols from gases implied choosing the most effective solid adsorbing material. Alumina gels, silica gels, activated carbons was tested for that purpose including those impregnated with Cu, Na, Ca and other salts. However, even the adsorption ability enhanced this way resulted in insufficiently high sulfur capacity [117]. In the contemporary industrial technology zeolite-containing adsorbents are of the maximal importance for gas sweetening [118].

Adsorption ability of zeolite adsorbents is caused by donor-acceptor interaction of polar molecules, for example thiols with oxygen atoms of silica-alumina framework and exchange cations neutralizing its negative charge [117]. Therefore, the higher cations content in zeolite and negative charge of its framework (for high-silica zeolites of types NaX, NaY and others), the higher its adsorption capacity. On saturating the zeolite sorbent layer, the breakthrough of the worst adsorbing component occurs; this means the end of the adsorption cycle, so the desorption is required being the stage of adsorbent regeneration.

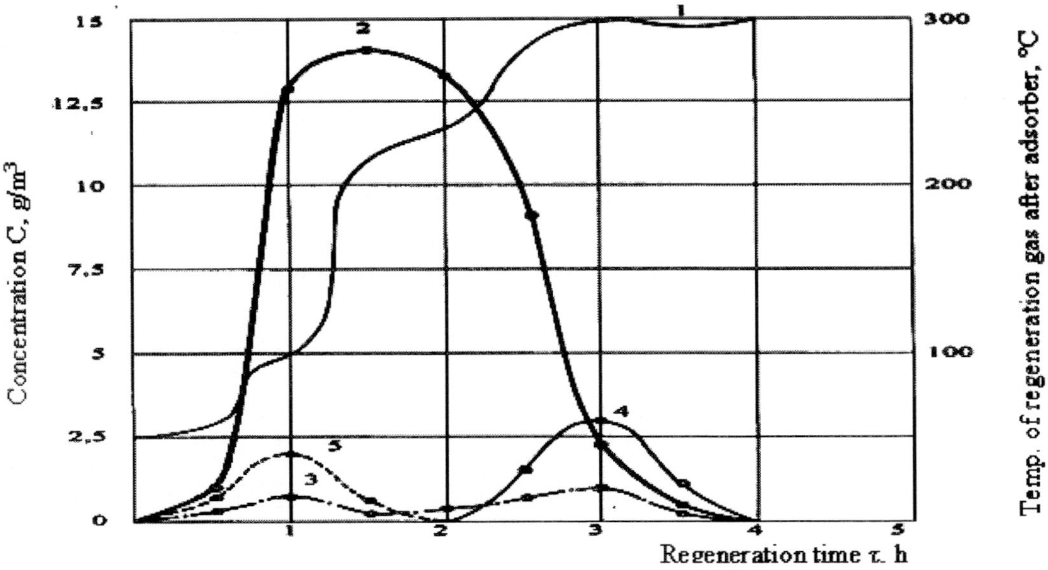

Figure 1.2. Thermodesorption of RSH (2), H2S (3), H2O (4) and C5+ hydrocarbons (5) on raising the temperature (1) in NaX zeolite layer.

Admixtures extracted from the raw gas are desorbed with the hot purified gas, as a rule [118]. Corresponding desorption curves observed (Figure 1.2) pointed to the difference in adsorption ability of H_2S, thiols, C_{5+} hydrocarbons and H_2O. Zeolite retains water most effectively; to remove water it is necessary to raise the temperature up to 290-320°C. And already at 150-170°C thiols, which are not desorbed yet, start cracking with formation of H_2S and alkenes, which form (under the action of zeolite adsorption sites enhancing their catalytic activity on temperature growth [119]) carbon deposits – the coke predecessors [117].

Zeolite sorbents accumulate carbon deposits during many alternating adsorption-thermodesorption cycles. These deposits deactivate zeolites [120]. Using combined silica-zeolite layer makes more efficiency [121]; the first for moving gas silica layer completely adsorb water, thiols and C_{5+} hydrocarbons, while the second zeolite layer adsorb H_2S. Thus, the thermodesorption with back flow of the hot purified gas takes out H_2S, C_{5+} hydrocarbons and thiols first, and then water at the temperature raised up to 200°C only. That almost excludes coking of zeolite adsorbent and prolongs its life time significantly.

Chemical methods for isolation of thiols from sulfurous raw hydrocarbon material are also known. In the beginning of XX century gases were sweetened by contacting with $Fe(OH)_3$ powder or iron-soda mass ($Fe(OH)_3$ + Na_2CO_3) with formation of $Fe(RS)_3$, subsequent oxidation into dialkylsulfides and regeneration of initial $Fe(OH)_3$ [110]. This technology possessed low productivity and not always provided the needed extent of gas purification.

Further fine purification was achieved using ZnO-based chemisorbents fixing thiols into ZnS. The process was carried out at 400-450°C and 2-4 MPa. Sometimes H_2 was introduced into reaction system for hydration of sulfurous compounds not fixed by chemisorbent. Later on, hydrofining of gases and liquid raw hydrocarbon material in the presence of Al-Co-Mo, Al-Ni-W etc. catalysts became the traditional process in oil and oil products processing [108]. Raw material of that type was also sweetened by treatment with small amount of H_2SO_4 (96-98%). In this case acid sludge formed not only from sulfurous but also from valuable aromatic compounds. In research practice selective precipitation of Hg or Ag thiolates was sometimes used [18].

Along with listed destructive methods of sweetening raw hydrocarbon material, in which thiols transformed into not valuable products, namely H_2S, elemental sulfur, acid sludge etc., mercaptans can undergo a lot of chemical reactions resulting in a wide range of products with various structure and properties (Sections 1.2 and 1.5).

1.5. Application of Natural Mercaptans (Alkylthiols) Mixtures for Chemicals Production

The problem of qualified application of sulfur-containing compounds from oils and gas condensates took on special significance in USSR in the middle fifties of the last century because of the impetuous development of sulfurous raw hydrocarbon material production and refining (at Ural-Volga fields), the part of which from the total volume became greater than 70% [122].

Investigation was conducted in two general directions.

- Application of sulfur-containing substances isolated from sulfurous raw hydrocarbon material, mainly mercaptans as chemicals useful for different areas of engineering, agriculture, medicine etc.
- Application of the same sulfur-containing substances of natural origin as a raw material for chemicals synthesis.

During the development of the first direction, a mixture of oil mercaptans turned out to possess slight foaming (floating) properties when dressing chalcopyrite ores, application of alkylsulfides, especially asymmetric dialkylsulfides, is much more effective [67]. For precious metals extraction from acid solutions alkylsulfides also occurred to be more effective than oil mercaptans [15]. Only C_{12}-alkylthiol synthesized from olefins and H_2S can be used for selective extractive isolation of Au and Ag, as well as for separation of Pd and Pt from them [123]. Testing oil mercaptans mixture as physiologically active substances have confined to determination of their toxicity level, which decreases on increasing their molecular mass [124], level of influence on central nervous system [56] and ascertainment of positive effect (when blended with oil sulfides) on animal's hair growth [55].

An important success was the isolation of mercaptans mixture with mean molecular mass about 190 from kerosene fraction of sulfurous oil from Ishimbayevskoye oil field and its application as a regulator of a molecular-mass distribution of polymers during butadiene-styrene resins synthesis [125]. A mixture of these mercaptans was brought to the routine formula consisting of monomers, isopropylbenzene peroxide and activator, then the low-temperature polymerization (–5-10°C) was carried out using conventional procedure [126]. Monomers conversion achieved nearly 60% for 18-19 h as for traditional stabilizer *tert*-C_{12}-thiol is applied.

For the preparation of the commercial resin SKS-30 with specified rigidity and viscosity the original formula needs to be added with 0.3 mass fractions of cheaper mercaptans into, that is only 1.5 times more than the amount of expensive *tert*-C_{12}-thiol prepared by synthetic method [125]. The similar results were further shown to be achieved with mercaptans mixtures isolated from corresponding distillates of Staroishimbayevskoye, Vvedenovskoye and Markovskoye oils of Bashkiria [127]. Despite the successful subsequent experimental tests of this development for production of SKS-30 resins at Stelitamak plant of synthetic resins [128], it was not applied for industrial manufacturing. The general reason is obviously a lack of industry-scale isolation of starting mercaptans.

The presence of raw material source (sulfurous natural gas and gas condensate) and the industrial basis for its processing allowed to organize production of C_2-C_5-thiols mixture that found an application as an odorant of industrial and domestic fuel gas and liquefied hydrocarbons [6, 114]. An odorant is brought into fuels in strictly limited amount (15-20 g/1000 m^3) for the purpose of opportune detection and prevention of gas leakage [80].

The following substances acted as odorants in Russia for a long period [2]: commercial-grade ethylmercaptan, sulfan – the concentrate of sulfur-containing organic compounds evaporating during sulfate cellulose pulping, containing up to 45% of highly toxic methylmercaptan and up to 50% of dimethylsulfide [129], as well as shale odorant (light fraction of shale gasoline including thiols and dimethylsulfide [80]). Gas-stock odorant (SPT odorant) turned out to be 1.35 times more effective and sufficiently cheaper [130], that provided its wide using. In 1991, nearly 4000 tons of SPT odorant was produced at Orenburg gas refinery (approximately 7% from mercaptans potentially isolated from gas condensate) [6]. Later on the odorant's quality has still improved, and production capacity has increased, its consumption, however, is limited by the volume of fuels to odorize.

Synthetic direction of the qualified application of natural thiols mixtures was readily being developed for the starting period. In the Institute of Organic Chemistry (IOC), USSR Academy of Sciences various vinyl derivatives were synthesized from narrow fraction of oil mercaptans [38, 52]; some of derivatives showed herbicidal and insecticidal activity rising with lengthen of *n*-alkyl group of thiol, the others were tested as floatation agents, metal extraction reagents, monomers for lacquering tars. Pesticidal activity was also revealed for products of thioalkylation of mucochloric acid with a mixture of oil mercaptans [131]. The same mixture was found useful as a therapeutic agent (antidote) against pesticide poisoning [15]; thiols' SH groups fixed the poison forming new C-S bonds and blocked the propagation of more complex and drastic processes.

Other investigators [25] subjected a mixture of oil C_{10}-C_{18}-alkylthiols to sulfochlorination and then saponifed the product to obtain detergents of not so high quality (the quality is close to the DS-RAS agent (refined alkylarylsulfonates), producing via sulfurization of oil fractions [132]). Being investigated was also the catalytic thiylation of *n*- and *iso*-olefins C_4-C_{10} with thiols mixture isolated from light distillate (b.p. 40-200°C) of Orenburg oil with formation of asymmetric dialkyldisulfides [37]. These dialkyldisulfides were extracted from the reaction mixture with H_2SO_4 aqueous solution. They proved to be effective extracting agents for precious metals [15, 17].

Even more active extracting agents for Ag, Au, Pd, Pt are ketosulfides [63-65] producing by thiomethylation of natural mercaptans mixture (Section 1.2). Synthesis of ketosulfides based on mercaptans isolated from Orenburg sulfurous gas condensate is developed with our participation [133]. Ketosulfides are also revealed [86] to be possibly used as more effective (in comparison with oil sulfides) growing accelerators for spring wheat, beet, potatoes, onions, peas and Lucerne increasing their crop capacity for 10-50% etc. The ability of ketosulfides to inhibit sulfurated hydrogen and common corrosion of steels [134].

The possibility of synthesis of *bis*-sulfides from oil mercaptans mixture [95] and their efficiency as stabilizers of polymethyl methacrylate thermal destruction [61] is studied. The stabilizing ability of *bis*-sulfides increases as *n*-alkyl group in starting thiol grows, and decreases with its branching. Iso-structured *bis*-sulfides are less stable towards thermal action that results in appearance of larger number of initiation sites even at their less concentration. However, studies listed and others did not bring investigators to development of industrial method for synthesis of chemicals based on alkylthiols mixtures of natural origin, primarily due to the lack of real source of raw material.

At the same time the considerable attention was paid to identify the role of mercaptans presenting in fuels and oils produced from sulfurous oils [17]. They were found [135], while presenting in jet engine fuel TS-1, that revealed catalytic action in this process. Negative influence of mercaptans does not observed only at concentrations lower than 0.007 mass. % at 100°C), and lower than 0.002% at 150°C. However *n*-C_{10} and *iso*-C_{12}-mercaptans (at the concentration 0.005-0.01 mass. %) inhibit oxidation of diesel oil from sulfurous oils [136].

The presence of mercaptans at low concentration in motor gasoline (0.0001 mass %) lowers its corrosive action towards metals, while at the concentration about 0.1 mass. % increases [137]. Addition of mercaptans into gasoline with antioxidant does not alter the induction period of its oxidation [138], but sharply decreases the efficiency of antiknock additive (lead tetraethyl) [139]. Mercaptans are found out [140] to promote oxidation of transformer oil, while removing sulfur-containing compounds from motor oil MT-16 enhance its corrosiveness for 2 times [141]. Considering these contradictory data it is recommended [17] to remove mercaptans from oil products as completely as possible, introducing, at the same time, synthetic additives including thiols derivatives, particularly SH-derivative of 2,6-di-*tert*-butyl-4-methylphenol (ionol) that is more active antioxidant than ionol itself (Table 1.9).

Table 1.9. Influence of ionol and its thiol derivative (at the concentration 0.05 mass. %) upon jet engine fuels during their thermal oxidation in the presence of copper (t = 160°C)

Additive	Fuel	Sediment, mg/100ml	Deposits on copper, g/m^2	Copper corrosion, g/m^2	Optical density in a layer 3.045 nm width
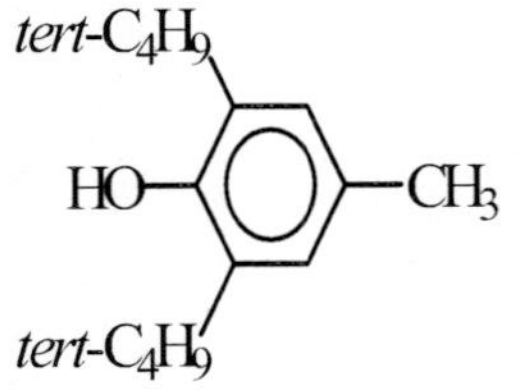	TS-1 RT-1 T-6	5.3 7.2 9.4	0.3 0.2 0.3	0.4 0.2 0.6	0.060 0.038 0.100
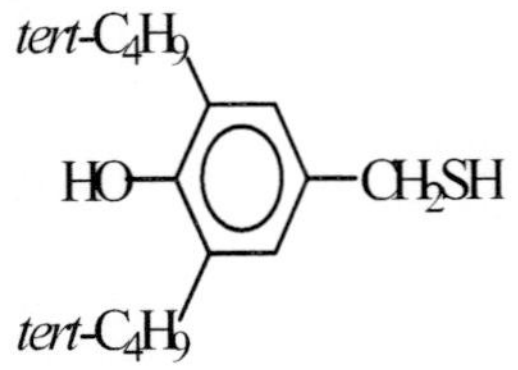	TS-1 RT T-6	2.6 2.8 2.6	0.1 0.1 0.2	0.2 0.1 0.2	0.003 0.008 0.009

These data along with many others listed above prove the expediency to develop investigations in the field of synthesis of oil products based on mixtures of natural thiols, in particular, the mixture of C_2-C_5-mercaptans, the actual industrial raw material source, isolated from Orenburg sulfurous gas condensate. On the analogy of earlier investigations [142, 143], the valuable products for low-tonnage chemistry can be produced containing fragments of *sym*-triazine and other heterocycles, aliphatic and aromatic amines, spatially hindered phenol, esters and possessing the wide spectrum of application properties (antioxidative, anticorrosive, physiological activity).

Chapter 2

Composition and Properties of C2–C5 Mercaptans Mixture of Natural Origin. Synthesis of ω-Alkylthioalkanols, Alkylthioacetic Acids and Their Derivatives on the Base of This Mixture

2.1. Composition of C2–C5 Mercaptans Mixture; Improvement of the Methods of Its Isolation Out of Orenburg Gas Condensate and of Its Properties as Industrial Odorant of Fuel Gas

As it was mentioned in the Section 1.4, at OGR there is the absorption plant isolating C_1–C_6-thiols mixture out of sulfurous gas condensate by water solution of sodium hydroxide [6, 114, 115]. This plant contains rectifying equipment for fractionating of isolated mixture of thiols [144]. In the same work they reported that degree of isolation of thiols mixture was about 60% in relation to their potential content in raw materials. Some researchers [15, 125] consider such low isolation degree as a consequence of oxidation of their noticeable part into dialkyldisulfides in water-alkaline solution, although reversibility of formation (hydrolysis) of sodium thiolates may turn out to be more significant reason [107]. The results of alkaline extraction of thiols with introduction into solutions of 10-20 mass % (0.9-6 mole/l) of additives (methanol, ethanol, cresol, etc.) increasing isolation degree of thiols testify in favor of prevailing contribution of the last factor [15, 125].

In connection with these facts we studied the efficiency of application of dialkylphthalates and mixtures of substituted pyrazoles and also pyrazolones known as widely distributed chemical products or semi-products in dyes production as minor additives which were able to rise the degree of thiols isolation by water solution of sodium hydroxide [145]. Laboratory experiments were carried out in standard glass separating funnels.

Initial sulfureous stable condensate containing 1.941 mass % of thiols C_1–C_6 was preliminary washed out (for removing of H_2S traces) by 20% water solution of diethanolamine (DEA) at temperature 20°C, then this solution was separated and treated by 10% water solution of NaOH; into NaOH solution the dibutylphthalate or pyrazole were added which concentrations were varied in the interval 0.0025-0.3 mole/l (we used their 20% solutions in *n*-propanol due to their limitary water solubility). When demercaptanization of hydrocarbon condensate was completed the isolated out of it thiols and condensate were analyzed by gas-liquid chromatography (GLC) [107, 146] with further calculation of their residual concentrations in condensate (C_c) and isolation degree (η). The results of experiments are presented in Tables 2.1 and 2.2.

It is obvious that tested additives for water-alkaline absorbent of thiols increase their total degree of isolation out of sulfur-containing gas condensate from 60-65% (with out additives) up to 95.4-99.0% first of all at the expense of extraction of additional amount of C_4–C_6-thiols. The maximum isolation degree of C_1–C_6-thiols mixture is reached when using dibutylphthalate additive in amount of 0.25 mole/l of water-alkaline solution (η = 95.4%). The higher value η=98.6% is reached with pyrazole additive taken in amount of 0.0025mole/l, i.e. pyrazole is more active additive (at its concentration 0.15mole/l the isolation degree of thiols mixture is 99%).

Table 2.1. The results of alkaline demercaptanization of sulfur-containing condensate with addition of dibutylphthalate

Thiols	Initial content of thiols, C_{in}, mass %	The content of dibutylphthalate in NaOH solution, mole/l							
		0,0025		0,126		0,25		0,30	
		C_c, mass %	η, %	C_c, mass %	η, %	C_c, mass %	η, %	C_c, mass %	η, %
C_1	0.006	0.0	100	0.0	100	0.0	100	0.0	100
C_2	0.563	0.0	100	0.0	100	0.0	100	0.0	100
C_3	0.848	0.0	100	0.0	100	0.0	100	0.0	100
C_4	0.397	0.071	82.1	0.055	86.1	0.043	89.2	0.043	89.2
C_5	0.081	0.036	55.6	0.031	61.7	0.028	65.4	0.028	65.4
C_6	0.046	0.023	50.0	0.020	56.5	0.019	58.6	0.019	58.6
Σ	1.941	0.130	93.3	0.106	94.5	0.090	95.4	0.090	95.4

Table 2.2. The results of alkaline demercaptanization of sulfur-containing gas condensate with addition of pyrazole

Thiols	Initial content of thiols, C_{in}, mass %	The content of pyrazole in NaOH solution, mole/l							
		0,0025		0,126		0,25		0,30	
		C_c, mass %	η, %	C_c, mass %	η, %	C_c, mass %	η, %	C_c, mass %	η, %
C_1	0.006	0.0	100	0.0	100	0.0	100	0.0	100
C_2	0.563	0.0	100	0.0	100	0.0	100	0.0	100
C_3	0.848	0.0	100	0.0	100	0.0	100	0.0	100
C_4	0.397	0.0	100	0.0	100	0.0	100	0.0	100
C_5	0.081	0.013	84.0	0.010	87.7	0.009	88.9	0.009	88.9
C_6	0.046	0.014	69.9	0.013	71.7	0.012	73.9	0.012	73.9
Σ	1.941	0.027	98.6	0.023	98.8	0.021	99.0	0.021	99.0

In connection with found activating effect caused by dibutylphthalate and pyrazole of extracting isolation of thiols out of sulfur-containing gas condensate we studied the nature of this effect. With this aim we established the character of interaction of mentioned additives and thiols in solution as it was described in work [147] and the shifts of SH groups signals in PMR-spectrum [12, 14]. We used the method of isomolecular series [148] with determination of refractive exponent at 20°C (n_D^{20}) and variability index $(n_D^{20})^2$ in dependence on compositions of solutions (Figures 2.1 and 2.2).

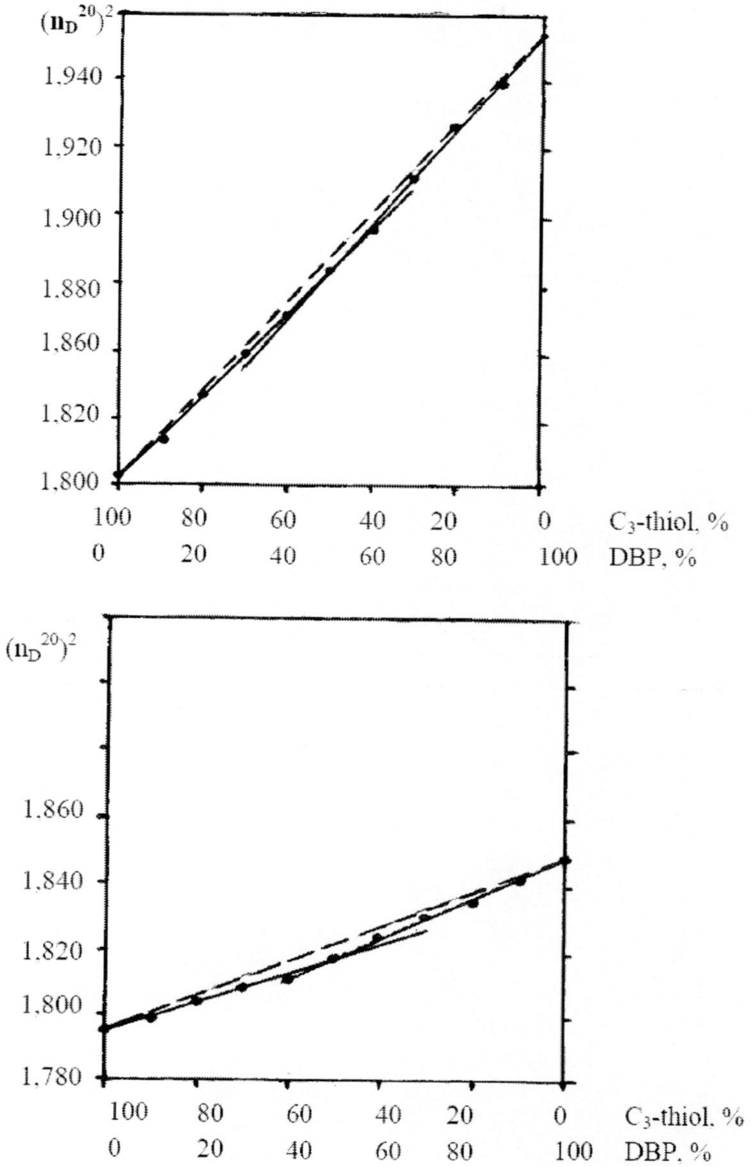

Figure 2.1. Change of (nD20)2 value in the isomolar series of dibutylphthalate-containing solutions of n-propanthiol (1.043mole/l) and n-hexanethiol (0.4955 mole/l) in methanol.

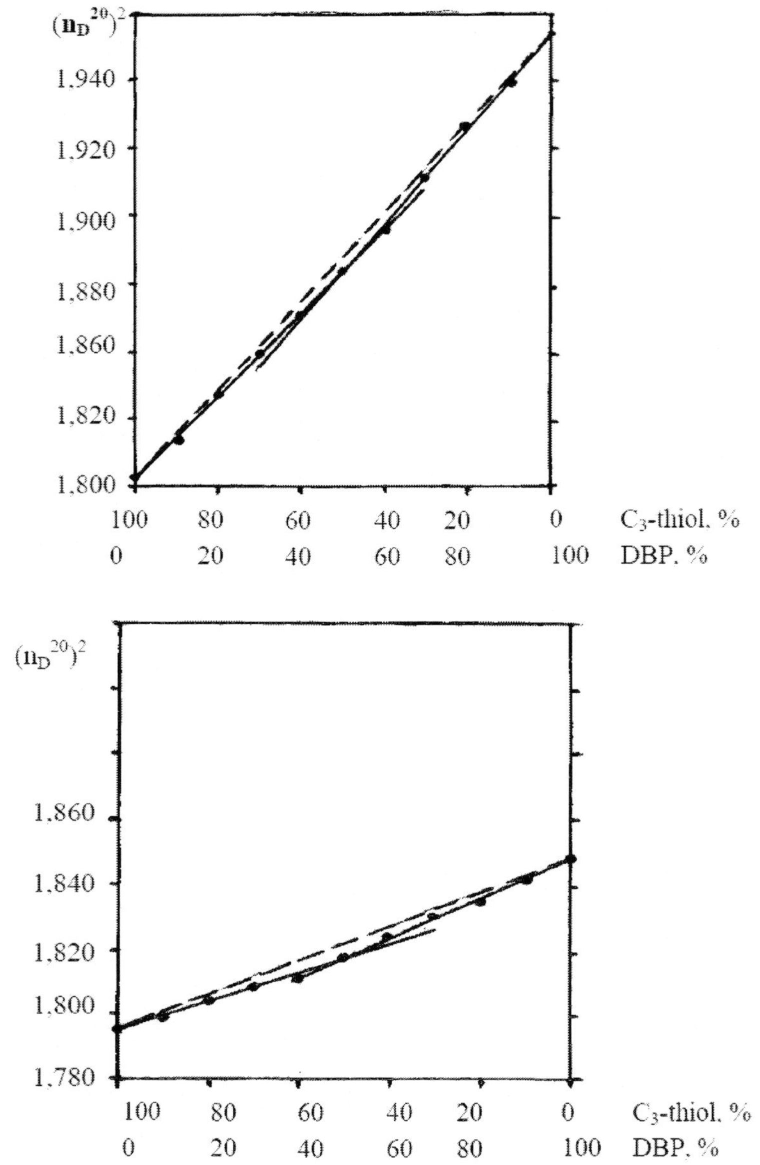

Figure 2.2. Change of (nD20)2 value in the isomolar series of pyrazole-containing solutions of n-propanthiol (1.043mole/l) and n-hexanethiol (0.4955 mole/l) in methanol.

The formation in solutions of equimolecular associates of thiols with dibutylphthalate and pyrazole corresponds to breaks on these dependences and by the value of specific deviation $\Delta(n_D^{20})^2$ (recalculation by concentration of solutions 1 mole/l) more active association f *n*-hexanethiol with pyrazole (specific deviation 0.0141) than with dibutylphthalate (specific deviation 0.00807) was established. *n*-Propanthiol weaker undergoes the association, but mentioned specific deviation for pyrazole is also higher (0.00719) than for dibutylphthalate (DBP) (0.00479).

Table 2.3. Chemical shifts of SH signals in PMR-spectra of solutions of *n*-propanthiol in methanol with addition of dibutylphthalate and pyrazole

Concentration of n-C_3-thiol, mole/l	δ_{SH}, m.p.	Concentration of DBP, mole/l	Concentration of n-C_3-thiol, mole/l	δ_{SH}, m.p.	Concentration of pyrazole, mole/l
1.043	6.92	0.0	1.043	6.92	0.0
0.834	7.09	0.209	0.834	7.22	0.209
0.626	7.18	0.417	0.626	7.39	0.417
0.417	7.29	0.626	0.417	7.58	0.626
0.209	7.42	0.834	0.209	7.81	0.834

Table 2.4. Chemical shifts of SH signals in PMR-spectra of solutions of *n*-hexanethiol in methanol with addition of dibutylphthalate and pyrazole

Concentration of n-C_3-thiol, mole/l	δ_{SH}, m.p.	Concentration of DBP, mole/l	Concentration of n-C_3-thiol, mole/l	δ_{SH}, m.p.	Concentration of pyrazole, mole/l
0.496	7.65	0.0	0.496	7.65	0.0
0.397	7.83	0.099	0.397	8.02	0.099
0.298	7.93	0.198	0.298	8.23	0.198
0.198	8.05	0.298	0.198	8.45	0.298
0.099	8.18	0.397	0.099	8.70	0.397

Analogous regularities were found with the use of PMR-spectroscopy (Tables 2.3 and 2.4). In PMR-spectra there is the shift of thiol group SH signal to the weak field as far as concentrations of dibutylphthalate and pyrazole are increased (Tables 2.3 and 2.4); as it is known this fact testifies to the formation of H-bond between thiol (donor) and additives (acceptors) activating its extraction. Moreover, carried out PMR-investigations confirm more essential acceptor effect of pyrazole in comparison with dibutylphthalate containing N and O atoms accordingly. And the weakest acceptor in studied solutions is obviously O atom in molecules of methanol solvent which is also able to form H-bond with thiol molecules [14].

Thus, we established that action of activating the thiols extraction dibutylphthalate and pyrazole added into NaOH solution consisted in the formation by them of donor-acceptor H-bonds with RSH molecules promoting the ionization of the last ones, especially of the thiols less dissociated in alkaline medium. Due to this fact the acidity of RSH is increased, the activity of $[RS^-]$ is risen, and correspondingly the value of K_{RSH} in equation (1.36) (Section 1.4) and also the constant of thiols distribution (K_q) in extracting agent (by the equation 1.34) are also increased. Similar suppositions were reported in the work [150], in which for increase of isolation efficiency of low-dissociated thiols out of hydrocarbon solution the water-organic solutions of alkali containing 20 vol. % of methanol and solvents of basic character − dimethyl formamide and dimethyl sulfoxide (60-70% of the volume of this mixture) were used. In our case the amount of activating additives (moreover inexpensive and profitable) introducing into alkaline extracting agent is significantly smaller.

On the base of found regularity we developed the novel recognized as inventions methods for the increase of thiols mixture yield during their isolation out of sulfur-containing gas condensate.

In accordance with inventors certificate of Russia 1027156 [151] the stable gas condensate containing 0.43 mass % of mercaptan sulfur and in two times larger volume of

10% water solution of NaOH including 0.5 mass % of dibutylphthalate and 15 mass % of methanol (or propanol) is fed into extractor. After agitation at 20°C for 10 minutes and aliquation (extraction is carried out in 3 steps) the demercaptanized condensate (the residual content of mercaptan sulfur is 0.04 mass %) is separated and out of water extract at 115°C thiols mixture is steamed. Their yield is 95% from potential content in condensate (without dibutylphthalate it is only 80%), thiols dissipation are reduced from 13.2 down to 3%.

In accordance with Russian patent 1077884 [152] the analogous treatment of gas condensate containing 0.83mass % of mercaptan sulfur is carried out with the use of water solution of NaOH including 0.5 mass % of pyrazole derivatives with general formula:

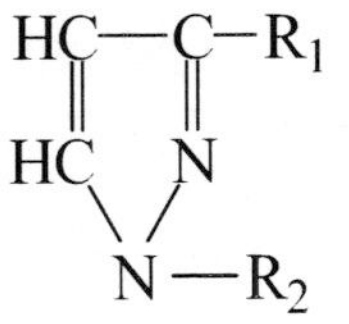

where $R_1 = H$, CH_3 or C_6H_5, $R_2 = H$ or C_6H_5 (waste product of azo dyes production), and 15% of ethanol. After agitation at 20°C for 10 minutes and aliquation (extraction is carried out in 1 step) the demercaptanized condensate (the residual content of mercaptan sulfur is 0.008 mass %) is separated and out of water extract at 115°C thiols mixture is steamed. Their yield is about 99% from potential content in condensate (without pyrazole it is 65%), dissipation are less than 1%.

In accordance with Russian patent 1416159 [153] natural sulfurous gas (90 vol. % of methane + 3 vol. % of ethane + 1 vol. % of propane + 6 vol. % of nitrogen) containing 50.1g/cm^3 of C_1–C_6 thiols mixture is treated in countercurrent absorber at -30°C and pressure 2MPa by C_4-C_6 thiols fraction in the amount of 0.9 l/m^3 of cleanable gas. The current of absorbent saturated by thiols undergoes eroding at atmospheric pressure with separating of absorbed hydrocarbons C_1–C_3 and then it is fractionated in rectifying column obtaining the mixture of C_1–C_4 and C_4–C_6 thiols which is reused as recirculating absorber. After demercaptanization natural gas contains 0.344g/m^3 of thiols, i.e. the yield of thiols is 99.3% from their potential content in raw material. In the last case thiols with larger molecular mass and correspondingly lower acidity serve as chemisorbents (at the expense of formation of H-bonds) of thiols mixture with wide fractional composition presenting in hydrocarbon crude.

The observing rise of isolation degree of thiols mixture out of natural hydrocarbon crude (sulfurous gas condensate and natural gas) is a great success in technology of its demercaptanization. However the fact how the rise of the yield of thiols mixture with wide fractional composition is reflected on its properties under its use as odorant remains unclear.

This thiols mixture selected at OGR in October, 2000 and boiling-off in the limits 7–135°C was analyzed by chromatography-mass spectrometry on "Finnigan MAT-1125" device in glass capillary column at the energy of ionizing emission 80eV (Table 2.5). It is obvious, that there are alkanes of regular structure (C_5-C_7 – 3.6 mass %, C_{15}-C_{19} – 2.3 mass %), aromatic hydrocarbons (benzene, toluene – 1.3 mass %), homologues and isomers of thiols (C_1-C_6 – 85.6 mass %) and also dialkyl(C_1-C_5)disulfides (7.2 mass %) in the composition of wide fraction of thiols.

Table 2.5. Composition of C_1-C_6 thiols mixture sample selected at OGR in October, 2000 (according to chromatography-mass spectrometry data)

No.	Components	Content, mass %	No.	Components	Content, mass %
1.	*n*-Pentane	1.0	16.	*n*-Butanethiol-1	3.5
2.	*n*-Hexane	1.5	17.	*n*-Butanethiol-2	17.5
3.	*n*-Heptane	0.7	18.	2-Methylpropanethiol-2	3.0
4.	Benzene	0.5	19.	*n*-Pentanethiol-1	0.6
5.	Toluene	0.9	20.	2-Methylbutanethiol-1	} 2.5
6.	Xylenes	0.3	21.	3-Methylbutanethiol-1	
7.	*n*-Pentadecane	1.0	22.	2-Methylpentanethiol-1	1.0
8.	*n*-Hexadecane	0.1	23.	Cyclopentanethiol	0.6
9.	*n*-Heptadecane	0.4	24.	Diethyldisulfide	1.3
10.	*n*-Octadecane	0.4	25.	Ethyl-*iso*-propyldifulfide	4.2
11.	*n*-Nonadecane	0.1	26.	Ethyl-*n*-propyldifulfide	0.7
12.	Methanethiol	0.5	27.	Isopropyl-*n*-propyldisulfide	0.7
13.	Ethanethiol	7.4	28.	Ethyl-*n*-butyldisulfide	0.9
14.	*n*-Propanethiol-1	15.5	29.	Methyl-*n*-pentyldisulfide	0.2
15.	*n*-Propanethiol-2	33.5	30.	*n*-Propyl-*n*-butyldisulfide	0.5

The mentioned thiols mixture undergoes rectification on the laboratory glass device with sampling of the fraction (83-86 mass %) boiling off in the limits 35–130°C (composition determined by gas-liquid chromatography is presented in Table 2.6) and of vat residue (6.5-7.9 mass %). Compositions of the same fractions which were obtained under rectification of thiols mixtures selected at OGR in March, 1995 and September, 1998 were also presented in Table 2.6 for comparing.

Table 2.6. Composition of thiols fraction boiling off in the limits 35-130°C (according to gas-liquid chromatography data obtained on LKhM-8 MD in the column 4000x3mm filled by celite bearer C-545+20% Reoplex 400 at 90°C and flow of gas-bearer of helium 3.6l/h)

No.	Components	Content in fraction (mass %) isolated out of thiols mixtures selected at OGR		
		October, 2000	September, 1998	March, 1995
1.	*n*-Hexane	2.0	1.8	1.3
2.	*n*-Heptane	0.5	0.9	0.8
3.	Benzene	1.0	0.9	0.8
4.	Toluene	1.0	1.1	0.7
5.	Ethanethiol	4.0	12.7	9.2
6.	*n*-Propanethiol-1	20.5	18.3	17.9
7.	*n*-Propanethiol-2	38.5	30.5	36.3
8.	2-Methylpropanethiol-2	3.5	1.9	3.7
9.	*n*-Butanethiol-2	20.0	19.8	15.9
10.	*n*-Butanethiol-1	4.5	6.6	5.2
11.	3-Methylbutanethiol-1	3.5	3.9	5.7
12.	*n*-Pentanethiol-1	1.0	1.6	2.5

There is some variation in the composition of one and the same fraction in dependence on the date of thiols mixture selection. Such changes were also observed in earlier works on studying of thiols composition selected at OGR [6, 107]. Obviously, composition of thiols mixture depends on the variation of parameters of technological regime of its extraction out of sulfur-containing gas condensate and steaming of thiols out of water-alkaline extract.

Distillate fractions of thiols mixture and initial mixture (the sample of 2000 year) were tested for intensity and stability of smell by GOST 22387.5-77 (State standard of Russia) by organoleptical study in odorizer by five-point scale. The results of tests are presented in Table 2.7. In accordance with presented data the maximum initial intensity of smell at odorization norm 20mg/m^3 of natural gas is characteristic for thiols fraction with boiling temperature 35−130°C (4 points), the lower one is observed for initial thiols mixture (3.7 points) and lass by 2.35 times (1.7 points) for fraction with boiling temperature higher than 130°C. In a minute the intensity of smell created by thiols fraction with boiling temperature 35−130°C is decreased down to 3.3 points (82.5% from the initial), in two minutes − down to 2.2 points (55%); for initial fraction these parameters are worse: in a minute − 2.8 points (75.6%), after 2 minutes − 1.8 points (48.6%); for fraction with boiling temperature higher than 130°C they are minimum: in a minute − 1.0 points (58.8%), after 2 minutes − 0.5 points (29.4%).

At lower odorization norm (10mg/m^3 of natural gas) the listed indexes are naturally reduced but in this case also the odorizating properties of fraction with boiling temperature 35−130°C remain the highest and for fraction with boiling temperature higher than 130°C − the lowest. So, high-boiling components of thiols mixture to a great extent are the ballast in the odorant structure.

However for odorant together with the most significant properties − intensity and stability of smell, the oxidation susceptibility and toxicity are also of great importance [80, 107]. As we mentioned in Section 1.2 thiols easily formed thiyl radicals under the action of oxidants and transformed into dialkylsulfides without smell at standard conditions. When natural gas containing odorant is transported by steel tubing the odorant interacts with oxidant − ferric oxides formed under the action of moisture on the internal walls of tubing with formation of dialkyldisulfides. As a result of this when gas is transported for long distances the smell of odorized gas is continuously reduced that requires the increased amount of odorant introduced into gas, i.e. the increase of its consumption is necessary.

Table 2.7. Intensity and stability of smell of thiols mixture isolated out of sulfur-containing gas condensate and its distillate fractions

Thiols mixture	Odorization norm of natural gas, mg/m^3	Intensity of smell (points) after:		
		0 min	1 min	2 min
Initial	20 10	3.7 (100%) 3.0 (100%)	2.8 (75.6%) 2.0 (66.7%)	1.8 (48.6%) 1.4 (46.6%)
Fraction with boiling temperature 35-130°C	20 10	4.0 (100%) 3.3 (100%)	3.3 (82.5%) 2.6 (78.8%)	2.2 (55.0%) 1.7 (51.5%)
Fraction with boiling temperature higher than 130°C	20 10	1.7 (100%0 1.0 (100%)	1.0 (58.8%) 0.5 (50.0%)	0.5 (29.4%) 0.2 (20.0%)

Various distillate fractions of thiols mixture containing larger or smaller amounts of 2-methylpropanethiol-2 (*tert*-butanethiol) possessing [5, 6, 136] the lowest among C_2-C_6 thiols oxidation susceptibility and revealing anti-oxidative action in relation to other thiols were used as test odorants. The data on oxidation stability of a number of samples of odorants (distillate fractions) of various structure determined by the developed simple technique (it is described in the Experimental part – see Appendix) are generalized in Table 2.8.

Table 2.8. Oxidation stability of odorant fractions containing propane- and butanethiols in various ratios

Properties of fractions	Fractions of odorant							
	1	2	3	4	5	6	7	8
Boiling temperature, °C	35-50	51-65	60-70	60-65	63-68	65-70	65-75	65-100
Composition of fractions, mass %: Ethanethiol	83.4	4.8	3.5	4.2	3.3	2.7	1.8	1.2
n-Propanethiol-1	5.8	14.7	36.6	22.7	27.9	35.8	36.5	29.4
n-Propanethiol-2	3.3	71.3	53.0	60.3	57.4	50.6	48.3	46.3
n-Butanethiol-1	0.2	0.2	0.3	0.2	0.2	0.2	0.3	6.9
2-Methylpropanethiol-2	1.5	3.0	5.8	7.0	4.7	4.1	3.9	2.8
3-Methylbutanethiol-1	0.1	0.1	0.1	0.1	0.1	0.1	0.2	0.9
n-Pentanethiol-1	0.0	0.0	0.0	0.0	0.0	0.0	0.1	0.2
Hydrocarbons	5.7	5.9	6.7	5.5	6.4	6.5	8.9	12.3
Yield of dialkylsulfides, %	16.5	6.7	5.8	5.0	6.1	6.5	6.8	12.6

Table 2.9. The properties of more stable to oxidation odorant in comparison with earlier known odorants

Odorants	Water solubility at 20°C, g/l	Dialkyl sulfides yield, %	Intensity of smell (points) after*)			Average-fatal dose for warm-blooded (mouses), mcg/g of mass
			0 min	1 min	2 min	
Stable to oxidation odorant (Patent 1125230, Russia)	1.1-1.3	7.0	4.2 (100%)	3.4 (81%)	2.3 (54.8%)	56.2
Known odorant – mixture of natural mercaptans (inventors certificate 654668, USSR)	1.50-1.52	15.1	3.7 (100%)	2.8 (75.7%)	1.8 (48.6%)	45.4
Ethylmercaptan (STU-12 No.10-232-62)	7.2-7.6	20.4	3.5 (100%)	2.4 (68.6%)	1.4 (40.0%)	15.0

Note: *) At odorization norm of natural gas 20mg/m^3.

In accordance with the data of Table 2.8 stability of narrow cuts of thiols is actually correlates with the content of *tert*-butanthiol in them: at its maximum content thiols mixture is oxidized with the lowest yield of dialkyldisulfides (5%). With the decrease of the part of *tert*-

butanthiol this yield is increased up to 16.5%; the lowest content of *tert*-butanthiol at which the yield of dialkylsulfides is acceptable (≤ 7%) is about 3.0 mass %.

On the base of carried out investigation the composition of more effective and stable to oxidation odorant including 3-7 mass % of *tert*-butanethiol (it may be superinduced if necessary), 15-20 mass % of *n*-propanthiol and *n*-propanthiol-2 (up to 100 mass %) was proposed and patented (Patent 1125230 [154] and 1214730[155], Russia, 1986). This odorant is boiled off in the limits 50-75°C, has the density 0.84 g/cm^3, and freezing temperature is about -110°C, steam tension at 30°C is in the limits 250-260 millimeters of mercury, flash point is 56-59°C, it contains38-42 mass % of mercaptan sulfur. It is also less in water solubility and toxicity for warm-blooded (average-fatal dose is higher) and characterized by higher stability of smell in comparison with earlier known odorants: the mixture of natural mercaptans (inventors certificate 654668, USSR, by MKI C 10 J 1/28, 1977) and synthetically produced standard ehtylmercaptan [80] (Table 2.9).

One more invention in this field (Patent 2105597, Russia, 2000 [156]) relates to improvement of the technology of thiols mixture isolation out of sulfur-containing hydrocarbon raw materials consisting in dewatering of isolated thiols on silica-gel and regeneration of this silica-gel by natural gas under heating.

Thus, carried out investigations allow edetermining the limits of variation of composition of industrially produced C_2-C_5 mercaptans mixture, proposing the methods of regulation of structure and properties of C_1-C_6 thiols mixture and also of its application as high-active odorant of natural gas. We showed that the main part (83-86 mass %) of mentioned thiols mixture is boiled off in the limits 35−130°C and is presented by close in reactivity thiols C_2-C_5 (content is 95.3-96.4 mass %) having alkyl groups of *n*- and *iso*-structure. This fraction may be recommended as both odorant and raw material for synthesis of a lot of sulfur-containing organic compounds possessing various useful applied properties analogously to those known chemicals considered in literature review (Section 1.3).

2.2. Synthesis and Properties of ω-C2-C5-Alkylthioalkanols, and C2-C5-Alkylthioacetic Acids and Their Derivatives

The literature data on reception and properties of mercaptans derivatives possessing high and versatile biological activity including those ones participating in biochemical processes (for example, glutathione, cysteine, dihydrolipoic acid) and also used as accelerants of rubbers vulcanization, stabilizers of polymer materials, combustion catalysts and oil dopes, etc. are summarized in the Section 1.3. Lability of bonds S−H and C−S, their ability to split under the action of both nucleophilic and electrophilic reagents, the presence of active reaction site containing sulfur of variable valency make thiols the valuable synthons widely used in thin organic synthesis [9, 10, 12, 79].

The most important direction in mercaptans chemistry is production of symmetric and asymmetric sulfides on their base by interaction of thiols and metals thiolates with alkyl- and arylhalogenides [11, 12]. We have already mentioned above (in Section 1.3) that on the base of fraction of natural mercaptans and elementary alkylchlorides the mixtures of dialkylsulfides with valuable applied properties were obtained.

2.2.1. Synthesis and Transformations of ω-Alkylthioalkanols

It is known that some ω-alkylthioalkanoles, for example 2-hydroxyethyl-, 3-hydroxypropyl- and 4-hydroxybutylsulfides are the important key substances for synthesis of various functional derivatives of sulfur-containing compounds [157-161] applied as effective plasticizers of polymer compositions [162].

Interaction of C_2-C_5 mercaptans mixture with 2-chloroethanole, 3-chloropropanole-1 and 4-chlorobutanol-1 was studied in presented work for the reception of such sulfides. We established that the best conditions for carrying out of these reactions are: the presence of equivalent amount of 20% solution of sodium hydroxide, temperature 50-60°C and duration of reaction mixture stirring 3-4 hours. At such conditions the desired ω-alkylthioalkanoles (Ia-c) are formed with the yield 80-86%:

$$NaOH + Cl(CH_2)_nOH \xrightarrow{NaOH} \underset{\textbf{Ia-c}}{AlkS\text{-}(CH_2)_nOH}$$

where Alk = alkyl C_2-C_5; n = 2 (a), 3 (b), 4 (c)

Oxidation of ω-alkylthioalkanoles (Ia-c) by H_2O_2 (30% concentration) in the mixture acetic acid–chloroform (1:1) at 20°C for 1-2 hours leads to ω-alkylsulfinylalkanoles (IIa-c) with the yield 72-77% [163]. By the other method of oxidative chlorination of ω-alkylthioalkanoles (Ia-c) by the mixture H_2O_2 (30%)–HCl (concentrated) according to the technique from [164] the ω-alkylsulfonylalkylchlorides (IIIa-c) are formed with essential yield (Table 2.10).

Table 2.10. Oxidative chlorination of ω-alkylthioalkanoles (Ia-c) and ω-alkylsulfinylalkanoles (IIa-c) by mixture H_2O_2 (30%)–HCl (concentrated)

Initial compound	n	Ia-c: H_2O_2 : HCl or IIa-c: H_2O_2 : HCl	Solvent	Yield of sulfones, %*)	
				III	IV
Ia	2	1 : 2 : 4	Acetonitrile	77	6
			Dioxane	75	12
			-	36	53
Ib	3	1 : 2 : 4	Dioxane	80	15
			THP	78	16
Ic	4	1 : 2 : 4	Acetonitrile	80	10
			THP	76	14
		1 : 4 : 8	Acetonitrile	70	18
			Methanol	45	50
			Trichloromethane	60	25
IIa	2	1 : 1 : 2	-	34	58
			Dioxane	82	10
IIb	3	1 : 1 : 2	-	33	56
			THP	78	13
IIc	4	1 : 1 : 2	-	30	64
		1 : 1 : 8	-	69	25
		1 : 1 : 2	Dioxane	72	23

Note: *) according to the data of gas-liquid chromatography of reaction products.

At that the corresponding ω-alkylsulfonylalkanoles (IVa-c; yield 10-16%) were separated out of reaction mixture as by-products. Oxidative chlorination was carried out at molar ratio sulfide Ia-c : H_2O_2 : HCl = 1 : 2-4 : 4-8 in the presence (or in the absence) of organic solvent in temperature interval from -10 up to +20°C.

The same sulfones IVa-c were obtained in high yield under oxidation of sulfides Ia-c or sulfoxides IIa-c by the action of 30% H_2O_2 (molar ratio Ia-c : H_2O_2 = 1 : 2.5-1 : 3) in acetic acid at 80-100°C for 1-1.5h:

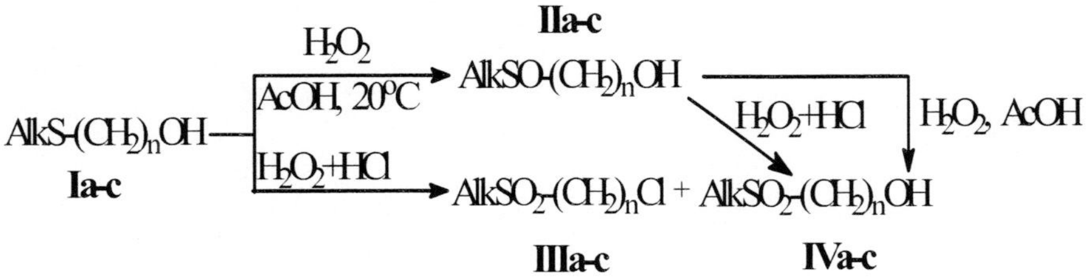

Alk = alkyl C_2–C_5; n = 2 (a), 3 (b), 4 (c)

Oxidative chlorination of ω-alkylsulfinylalkanoles (IIa-c) by mixture H_2O_2 (30%)–HCl (concentrated) at the conditions mentioned above also leads to mixtures of corresponding cloralkylsulfones IIIa-c and ω-alkylsulfonylalkanoles IVa-c in the ratio about 1 : 2 (Table 2.10). It is significant that with the rise of distance between HO-group and sulfur atom in sulfides Ia-c and sulfoxides IIa-c the yield of sulfones IVa-c is increased. Obviously, it may be explained by heterogeneous conditions of the process catalyzed by HCl and competitive oxidation by HClO (or H_2O_2) of initial compounds into sulfones IVa-c [164].

The rise of the content of H_2O_2 in reaction medium slightly changes the ratio of resulted sulfones IIIa-c and IVa-c. On the contrary the rise of HCl content significantly influences on process leading to the yield of chlorine-derivatives up to 70-80%. Obviously the rise of molar content of HCl in reaction medium shifts the equilibrium in system H_2O_2–HCl to formation of free chlorine that inhibits competitive reaction of oxidation of sulfides Ia-c or sulfoxides IIa-c into ω-alkyl-sulfonylalkanoles [160, 163].

It was found that nature of organic solvent influences on the yield of chlorides IIIa-c: in methanol medium the yield of 4-alkylthiobutylchlorides (IIIb) is 45% and in homogeneous solution of dioxane, acetonitrile or tetrahydrofuran the yield of chlorides IIIa-c is increased up to 72-80%. At that the variation of the order of reagents addition – introduction of sulfide Ia-c (or sulfoxide IIa-c) into mixture H_2O_2–HCl or H_2O_2 into mixture sulfide Ia-c (or sulfoxide IIa-c)–HCl, and also temperature regime (from -15 up to +20°C) slightly influences on the yield of ω-alkylsulfonylalkylchlorides IIIa-c. They were obtained with the yield 75-80% at molar ratios sulfides Ia-c : H_2O_2 HCl = 1 : 2 : 4 or sulfoxide IIa-c : H_2O_2 : HCl = 1 : 1 : 2 in the solution of acetonitrile or dioxane at 0-5°C.

Carried out by us oxidative bromination of sulfides Ia-c or sulfoxides IIa-c under the action of mixture H_2O_2 (30%)–HBr (40%) proceeds analogously to oxidative chlorination and is resulted in the mixture of ω-alkylsulfonylalkylbromides Va-c and ω-alkylsulfonylalkanoles IVa-c:

$$\left.\begin{array}{l} AlkS\text{-}(CH_2)_nOH \\ \textbf{Ia-c} \\ \\ AlkSO\text{-}(CH_2)_nOH \\ \textbf{IIa-c} \end{array}\right] \xrightarrow{H_2O_2+HBr} \underset{\textbf{Va-c}}{AlkSO_2\text{-}(CH_2)_nBr} + \underset{\textbf{IVa-c}}{AlkSO_2\text{-}(CH_2)_nOH}$$

where Alk = alkyl C_2-C_5; n = 2 (a), 3 (b), 4 (c)

The particularity of this method is in the fact that the yield of brom-derivatives Va-c depends on the order of reagents introduction and their molar ratio. At the same time essential differences between reaction carrying out with or without organic solvent were not observed. With the help of model experiments with the use of β-alkylthioethanoles Ia mixture we showed that the rise of molar ratio HBr : sulfides Ia-c higher than 2 in reaction medium and addition of sulfides to H_2O_2–HBr led to the decrease of bromsulfones Va yield. In this case the best results were obtained under addition of H_2O_2 to the mixture sulfide Ia–HBr.

Under oxidative bromination of β-alkylthioethanoles Ia and β-alkylsulfinylethanoles IIa the optimal condition of reception of β-alkylsulfonylethylbromides Va is the molar ratio sulfides Ia : H_2O_2 : HBr = 1 : 2 : 1.5-2 (or sulfoxides IIa : H_2O_2 : HBr = 1 : 1 : 1.5-2). At these conditions one is succeeded in selective direct of the process to the reception of bromides Va and in significant decreasing (down to 1-3%) of the amount of side products such as sulfones IVa (Table 2.11).

We established under oxidative bromination of γ-alkylthiopropanols Ib and δ-alkylthiobutanols Ic or sulfoxides IIb, c that the optimal molar ratio is sulfides Ib, c : H_2O_2 : HBr = 1 : 2 : 4 or sulfoxides IIb, c : H_2O_2 : HBr = 1 : 1 : 2-4, at which the yield of corresponding bromalkylsulfones Vb, c is 70-80%.

It is known [10, 165] that ω-alkylthioalkanoles serve as suitable synthons under production of various functionally-substituted sulfur-organic compounds.

Table 2.11. Oxidative bromination of ω-alkylthioalkanoles (Ia-c) and ω-alkylsulfinylalkanoles (IIa-c) by mixture H_2O_2 (30%)–HBr (40%)

Initial compound	n	Ia-c: H_2O_2 : HCl or IIa-c: H_2O_2 : HCl	Solvent	Yield of sulfones, %*⁾	
				V	IV
Ia	2	1 : 2 : 1.5	-	77	2
		1 : 2 : 4	Dioxane	65	5
Ib	3	1 : 2 : 4	Dioxane	80	4
			-	83	10
			Acetonitrile	75	5
IIa	2	1 : 1 : 2	-	81	12
			Dioxane	72	8
IIb	3	1 : 1 : 2	-	84	8
			Acetonitrile	73	9
IIc	4	1 : 1 : 2	-	52	35
		1 : 1 : 8	-	68	10
		1 : 1 : 2	Dioxane	81	11

Note: *⁾ according to the data of gas-liquid chromatography of reaction products.

In our work we studied reactions of alcohols Ia-c alkylation by chloromethyl ether and also their amino-methylation under the action of formaldehyde and secondary amines. So, under the interaction of equimolecular amounts of ω-alkylthioalkanoles (Ia-c) and chloromethylether in waterless benzene at 20°C with further ageing of reaction mixture at 40°C (for 3-4 hours) the mixtures of α-(methoxymethoxy)-ω-alkylthioalkanes (VIa-c) were synthesized with the yield 70-76% (Table 2.12). Tertiary amines − triethylamine or N,N-dimethylaniline were used as HCl acceptor.

$$\underset{\mathbf{Ia\text{-}c}}{AlkS\text{-}(CH_2)_n\text{-}OH} \xrightarrow[-HCl]{ClCH_2OMe} \underset{\mathbf{VIa\text{-}c}}{AlkS\text{-}(CH_2)_n\text{-}OCH_2OMe}$$

$$\underset{\mathbf{Ia\text{-}c}}{AlkS\text{-}(CH_2)_n\text{-}OH} \xrightarrow{(CH_2O)_x + R^1RNH} \underset{\mathbf{VIIa\text{-}c}}{AlkS\text{-}(CH_2)_n\text{-}OCH_2NRR^1}$$

where Alk = alkyl C_2-C_5; n = 2 (Ia, VIa, VIIa-i), 3 (Ib, VIb, VIIj-n), 4 (Ic, VIc, VIIo-r); $R^1RN = (C_2H_5)_2N$ (VIIa, i, n), $(n\text{-}C_4H_9)_2N$ (VIIb), $(n\text{-}C_8H_{17})_2N$ (VIIc), $n\text{-}C_4H_9NCH_3$ (VIId), $n\text{-}C_{18}H_{37}NCH_3$ (VIIe, j), 4-HO-3,5-$(tert\text{-}C_4H_9)_2C_6H_2CH_2NCH_2Ph$ (VIIf, k, o), morpholino (VIIg, l, p), piperidino (VIIh, m, q)

Reaction between equimolar amounts of ω-alkylthioalkanoles (Ia-c), paraform and secondary amines in waterless benzene at 20°C (for 1 hour) with further ageing of reaction mixtures at 50-60°C for 4-6 hours leads to high yield (Table 2.13) of amino-esters − N,N-disubstituted α-(aminomethoxy)-ω-alkylthioalkanes (VIIa-c).

Physical-chemical characteristics of synthesized compounds Ia-c, IIa-c, IIIa-c, IVa-c, Va-c, VIa-c and VIIa-c are presented in Tables 2.12 and 2.13. It is obvious that they are boiled off (or melt) in essentially narrower temperature interval in comparison with the initial mixture of mercaptans (95-130°C). This fact is probably conditioned by both transformations of mercaptans into sulfides (thioesters) which C_2–C_5-homologues have smaller diapason of physical properties change [9, 15, 17], and also by the decrease of mass part of sulfide fragments in the molecules of obtained by us mixtures of S, N, O–containing compounds. Actually, with the rise of molecular mass in the series, for example such as Ia – Ib – Ic, VIIa – VIIb – VIIc and so on the gradual decrease of interval of boiling temperatures of these compounds is observed (the diapasons of boiling temperatures correspondingly are 28, 27, 26 and 22, 21, 20°C at reduced pressure, and at standard – they are by 30% larger).

Structure of synthesized mixtures of one-type compounds is confirmed by IR- and PMR-spectroscopies data, and compositions are conformed by elementary analysis. In IR-spectra of Ia-c, IIa-c and IVa-c there are broad bands in the region 3480-3240cm^{-1} characteristic for stretching vibrations of H–O bonds in associated hydroxyl groups [166]. The number of absorption bands of average intensity in the interval 1335-1280cm^{-1} relates to stretching vibrations of C–O bond [167] in primary alcohols Ia-c, IIa-c, IVa-c. In spectra IIIa-c, Va-c, VIa-c and VIIa-c absorption in the region 3480-3200cm^{-1} is absent.

Table 2.12. Constants, yields and data of elementary analysis of ω-alkylthioalkanoles (Ia-c), ω-alkylsulfinylalkanoles (IIa-c), ω-alkylsulfonylalkylchlorides (IIIa-c), ω-alkylsulfonylalkanoles (IVa-c) and ω-alkylsulfonylbromides (Va-c) Alk – S(O)x–(CH2)n–Z

No.	n	x	Z	Yield, %	$T_{boil.}$, °C (millimeters of mercury) or $T_{melt.}$, °C (solvent for crystallization)	n_D^{20}	d_{20}^{4}	S, mass %
Ia	2	0	OH	86	38-66 (2)	1.4796	1.1014	25.12
Ib	3	0	OH	82	50-77 (2)	1.4811	1.1041	23.08
Ic	4	0	OH	80	54-80 (1)	1.4819	1.1062	20.92
IIa	2	1	OH	72	oil*⁾	1.4754	1.0912	22.28
IIb	3	1	OH	77	oil *⁾	1.4782	1.0854	19.85
IIc	4	1	OH	75	34-39*⁾	-	-	19.10
IIIa	2	2	C1	73	34-41 (*n*-hexane)	-	-	17.37
IIIb	3	2	C1	67	oil*⁾	1.4905	1.1501	17.08
IIIc	4	2	C1	65	40-46 ($CC1_4$/*n*-hexane, 2:1)	-	-	14.86
IVa	2	2	OH	88	32-37 (*n*-hexane)	-	-	19.17
IVb	3	2	OH	83	44-48 (*n*-hexane)	-	-	19.08
IVc	4	2	OH	86	50-56 (*n*-hexane)	-	-	16.12
Va	2	2	Br	81	oil*⁾	1.5108	1.2304	14.84
Vb	3	2	Br	84	51-56 ($CC1_4$/*n*-hexane, 1:1)	-	-	13.04
Vc	4	2	Br	76	40-47 (*n*-hexane)	-	-	13.27

Note: *⁾ Substances were purified by column chromatography on silicagel.

Table 2.13. Constants, yields and data of elementary analysis of α-(methoxymethoxy)- (VIa-c) and N,N-disubstituted α-(aminomethoxy)-ω-alkylthioalkanoles (VIIa-c). Alk – $S(CH_2)_n–OCH_2–Z$

No	n	Z	Yield, %	$T_{boil.}$, °C (millimeters of mercury) or $T_{melt.}$, °C (solvent for crystallization)	n_D^{20}	d_{20}^4	S, mass %
VIa	2	CH_3O-	76	43-68 (2)	1.4688	1.1109	18.94
VIb	3	CH_3O -	70	48-74 (2)	1.4693	1.1114	17.72
VIc	4	CH_3O-	72	52-76 (1)	1.1679	1.0978	16.24
VIIa	2	$(C_2H_5)_2N$-	74	67-90 (1)	1.4954	1.1084	14.20
VIIb	2	$(n\text{-}C_4H_9)_2N$-	70	98-120 (1)	1.4972	1.1198	11.64
VIIc	2	$(n\text{-}C_8H_{17})_2N$-	73	43-48*)	-	-	8.04
VIId	2	$n\text{-}C_4H_9(CH_3)N$-	80	46-98 (2)	1.4887	1.1037	13.87
VIIe	2	$n\text{-}C_{18}H_{37}(CH_3)N$-	77	36-42 (acetone)	-	-	7.84
VIIf	2	4-HO-3,5-(*tert*-$C_4H_9)_2C_6H_2CH_2(CH_2Ph)N$-	68	55-62*)	-	-	7.04
VIIg	2	morpholino-	75	90-113 (2)	1.4915	1.1120	13.52
VIIh	2	piperidino-	82	102-125 (2)	1.4945	1.1205	13.88
VIIi	3	$(C_2H_5)_2N$-	72	82-103 (2)	1.4710	0.9972	14.62
VIIj	3	$n\text{-}C_{18}H_{37}(CH_3)N$-	76	45-52 (acetone)	-	-	6.86
VIIk	3	4-HO-3,5-(*tert*-$C_4H_9)_2$-$C_6H_2CH_2(CH_2Ph)N$-	70	63-70*)	-	-	7.08
VIIl	3	morpholino-	71	103-124 (1)	1.4986	1.0085	13.71
VIIm	3	piperidino-	67	107-128 (1)	1.5008	1.0212	12.18
VIIn	4	$(C_2H_5)_2N$-	65	87-109 (1)	1.5064	1.0432	13.66
VIIo	4	4-HO-3,5-(*tert*-$C_4H_9)_2$-$C_6H_2CH_2(CH_2Ph)N$-	62	72-80*)	-	-	6.07
VIIp	4	morpholino-	70	113-133 (2)	1.5087	1.0227	12.18
VIIq	4	piperidino-	74	122-141 (2)	1.5110	1.0564	13.05

Note: *) Substances were purified by column chromatography on silicagel.

In spectra of IIa-c intensive maximums of absorption at 1050-1035cm^{-1} characteristic for vibration of S=O group in sulfoxides are observed [167]. Spectra of sulfones IIIa-c, IVa-c, Va-c contain strong absorption bands in the regions 1320-1300 and 1150-1130cm^{-1} which are conditioned by asymmetric and symmetric stretching vibrations of SO_2 group [167].

In PMR spectra the same signals of protons of CH_3-groups represent dublets and triplets at 0.88-1.12m.p. Complex multiplets in the interval 1.24-1.73 m.p. should be assigned to protons of $-CH_2-$ units, in the interval 1.70-1.92m.p. – to protons of CH_3-groups, and signals shifted to weak field (2.58-3.93 m.p.) – to protons of CH_2- and CH– groups bonded with sulfur atom. In spectra of alcohols Ia-c, IIa-c and IVa-c the signals of hydroxyl protons are presented by broadened signals at 4.95-5.30m.p. Asymmetric triplets in the interval 3.55-3.96m.p. relate to protons of $-CH_2O-$ groups [167]. In spectra of compounds VIa-c and VIIa-c protons of methylene groups $-OCH-_2O-$ and OCH_2N< are characterized by singlet signals at 4.15-4.42m.p.

Table 2.14. Individual compositions of mixtures of synthesized ω-alkylthioalkanoles derivatives (according to chromatography-mass spectrometry data)

Alk	The content of component in mixture of compounds, %							
	Ib	IIa	IIIc	IVa	Vb	VIb	VIIa	VIId
C_2H_5	3.0	3.5	3.0	2.9	3.3	3.5	2.5	2.8
n-C_3H_7	24.0	21.0	22.5	22.9	19.8	20.5	23.5	21.7
iso-C_3H_7	39.5	40.5	39.0	39.8	41.5	40.0	39.5	39.9
n-C_4H_9	4.8	5.0	5.0	5.2	5.3	5.5	5.0	5.1
iso-C_4H_9	23.2	24.6	24.5	23.7	23.5	23.5	23.0	24.0
n-C_5H_{11}	1.5	1.9	1.8	1.7	1.9	2.0	2.5	2.1
iso-C_5H_{11}	4.0	4.0	4.2	3.8	4.7	5.0	5.0	4.4

The contents of individual components in mixture of reactions' products were determined for some synthesized substances with the help of chromatography-mass spectrometry (Table 2.14). According to these data the ratio of substituents C_2-C_5 (including isomers) in ω-alkylthioalaknoles and their derivatives is close (with consideration of experimental accuracy) to the content of homologues and isomers of thiols in initial mixture of mercaptans. Obviously reactivities of mercaptans mixtures components in presented chemical transformations are approximately equal.

2.2.2. Synthesis and Transformations of C2–C5-Alkylthioacetic Acids and Their Functional Derivatives

Alkylthioacetic acids and their derivatives are of independent interest as stabilizers of polymer materials [8, 168] and lubricating oils [72] and also are the suitable semi-products in the synthesis of various sulfur-organic compounds of aliphatic series and nitrogen-containing hetero-cycles with alkylthiomethyl groups [10, 79].

First of all we developed preparative methods of production of mixture of alkylthioacetic acids VIII and their basic functional derivatives – chloroanhydrides IX, ethers Xa-h, amides XIa, n-substituted amides XIb-g and hydrazides XII on the beas of mercaptans C_2-C_5 mixture.

Alkylthioacetic acids VIII (with the yiled 80-86%) were synthesized by boiling of mentioned mercaptans mixture with chloroacetic acid (for 2 hours) in water solution of alkali (mole ratio 1 : 1 : 2) and further neutralization by concentrated HCl at 0°C and corresponding chloroanhydrides IX of acids with the yield 78-82% were obtained by boiling with double excess of thionylchloride in waterless benzene (for 3-4 hours).

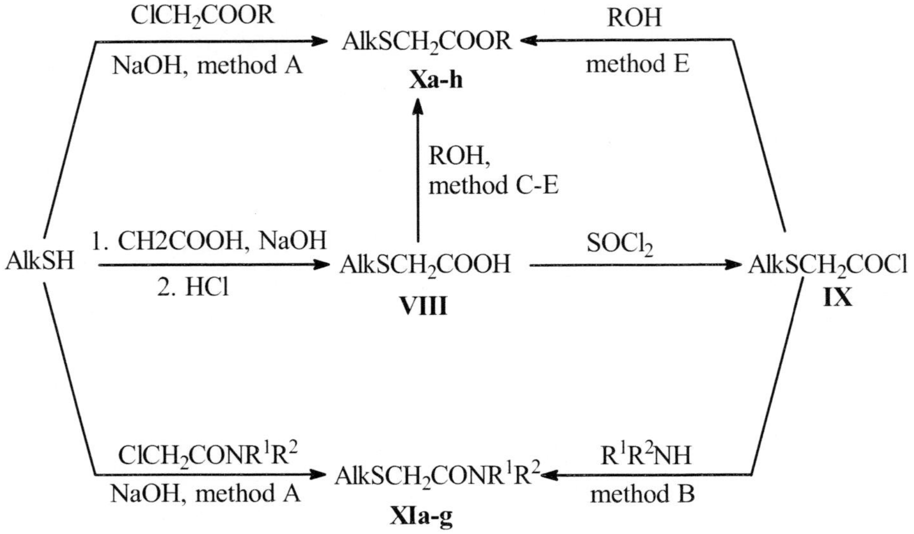

Alk = alkyl C_2-C_5;

X: R = CH_3 (a), C_2H_5 (b), *н*-C_4H_9 (c), *н*-C_8H_{17} (d), tetrahydrofurfuryl (e), Ph (f), 4-ClC_6H_4 (g), 4-HO-3,5-(*tert*-C_4H_9)$_2C_6H_2$ (h);

XI: R^1R^2N = NH_2 (a), $(C_2H_5)_2N$ (b), н-$C_8H_{17}NH$ (c), PhNH (d), 4-$CH_3OC_6H_4NH$ (e), thiazolyl-2-amino (f), morpholino (g).

We carried out the search of the most suitable preparative methods of synthesis of ethers of alkylthioacetic acids Xa-h including alkylation of C_2–C_5 mercaptans mixture by ethers of choloroacetic acid (method A), esterification of acids mixture VIII by alcohols in the presence of concentrated H_2SO_4 (method B) or other condensing agents (methods C–E) and also acylation of alcohols and phenols by chloroanhydrides (method E). The comparative results of tests of methods are presented in Table 2.15.

So, acylation of mercaptans mixture by corresponding ethers of chloroacetic acid was carried out in aqueous alcohol in the presence of alakli (molar ratio 1 : 1 : 1.2) at 20°C with further heating of reaction mixture up to 50-60°C for 2 hours (method A). Ethyl (Xb) and butyl (Xc) ethers of thioacetic acid were obtained with the yield 53-57% under boiling (for 5-6 hours) of acids VIII mixture with ethanol or butanol-1 (molar ratio 1 : 5) in the presence of 5-10 vol. % of H_2SO_4 (method B). However in the case of octanol-1 and tetrahydrofurfuryl alcohol the yield of ethers Xd, e did not exceed 25-30%.

Among known reagents [168] applied for activating of synthesis of esters of carbonic acids we tested the most available of them: trimethylchlorosilane [169], methanesulfochloride + triethylamine [170] and diphenylchlorophosphate + dimethylformamide (DMFA) [171]. It turned out that interaction of acids mixture VIII with methanol, ethanol, butanol-1 and tetrahydrofurfuryl alcohol in the presence of trimethylchlorosilane (molar ratio of these reagents 1 : 4 : 4) in waterless tetrahydrofuran (THF) at 30-35°C (for 12 hours) led to ethers Xa, b, c, e (method B) with the yield 86-90% (Table 2.15).

Esterification of acids VIII mixture by alcohols and phenols in the presence of methanesulfochloride and triethylamine (molar ratio of reagents 1 : 4 : 1 : 2) in dichloromethane (0°C; 25-30°C, 4 hours) leads to formation of ethers Xa, b, e, g with the yield 84-90% (method D). Reactive complex [$(PhO)_2POCl$]–dimethylformamide is formed at 0°C in the reaction of waterless dimethylformamide with diphenylchlorophosphate; under the treatment of this complex in dichloromethane (20°C, 2 hours) by acids VIII mixture, alcohol or phenol and pyridine (molar ratio 1 : 2 : 4) the ethers Xb, d, f, g with the yield 88-94% were obtained (method E).

The most convenient preparative method of synthesis of ethers Xa-h is obviously the acylation of alcohols and phenols by chloroanhydrides IX in the presence of pyridine (molar ratio of reagents 1 : 1 : 2) in dioxane (method F). In this case at temperature 20-30°C for 2-3 hours the yield of final ethers Xa-h reaches 76-89% (Table 2.15).

The corresponding amides of alkylthioacetic acids XIa and XIb-g (method A) with the yield 65-78% were synthesized by reaction of C_2–C_5 mercaptans mixture with chloroacetoamide or N-substituted chloroacetoamides in aqueous alcohol (boiling for 3 hours) in the presence of equivalent amount of alakli (Table 2.16). We showed that it was advisable to synthesize these amides by acylation of corresponding amines by chloroanhydrides IX under boiling (for 3 hours) in waterless benzene in the presence of triethylamine (molar ratio 1 : 1 : 1.5). The yield of N-substituted amides XIb-g obtained by method B is practically the same (76-88%) as it was when we used method F.

Table Hydrazides of alkylthioacetic acids (XII) with the yield 70-75% were synthesized by boiling of methyl ethers Xa with 85% hydrazine hydrate (molar ratio 1 : 1.5 : 4.0) in propanol-2:

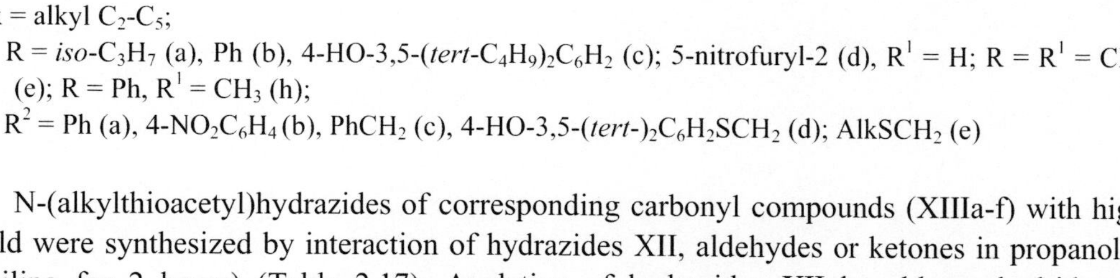

Alk = alkyl C_2-C_5;

13: R = *iso*-C_3H_7 (a), Ph (b), 4-HO-3,5-(*tert*-$C_4H_9)_2C_6H_2$ (c); 5-nitrofuryl-2 (d), R^1 = H; R = R^1 = CH_3 (e); R = Ph, R^1 = CH_3 (h);

14: R^2 = Ph (a), 4-$NO_2C_6H_4$ (b), $PhCH_2$ (c), 4-HO-3,5-(*tert*-)$_2C_6H_2SCH_2$ (d); $AlkSCH_2$ (e)

N-(alkylthioacetyl)hydrazides of corresponding carbonyl compounds (XIIIa-f) with high yield were synthesized by interaction of hydrazides XII, aldehydes or ketones in propanol-2 (boiling for 2 hours) (Table 2.17). Acylation of hydrazides XII by chloroanhydrides of various acids in pyridine (boiling for 2 hours) leads to N-(alkylthioacetyl)-N-acylhydrazides XIVa-e with the yield 72-85% (Table 2.18).

Table 2.15. Constants, yield and data of elementary analysis of ethers of alkylthioacetic acids (Xa-h) AlkSCH$_2$COOR

Compound	R	Yield, % (preparative method)	$T_{boil.}$, °C (millimeters of mercury) or $T_{melt.}$, °C (solvent for crystallization)	n_D^{20}	d_4^{20}, g/cm^3	S, %
Xa	CH_3-	74 (A), 85 (C), 86 (D), 80 (F)	41-66 (2)	1.4624	1.097	20.08
Xb	C_2H_5-	77 (A), 53 (B), 90 (C), 84 (D), 88 (E), 85 (F)	51-75 (2)	1.4667	1.105	19.21
Xc	*н*-C_4H_9-	72 (A), 57 (B), 88 (C), 78 (F)	60-84 (2)	1.4684	1.118	16.74
Xd	*н*-C_8H_{17}-	68 (A), 90 (D), 78 (F)	90-112 (2)	1.4702	1.138	13.94
Xe	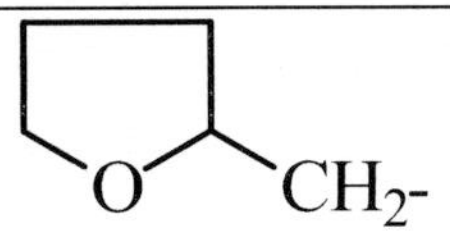	72 (A), 90 (C), 86 (D), 81 (F)	80-103 (2)	1.4638	1.105	13.71
Xf	Ph-	83 (A), 94 (E), 87 (F)	48-54 (ethanol-H_2O, 1:1)	-	-	14.18
Xg	4-ClC_6H_4-	79 (A), 88 (D), 90 (E), 89 (F)	64-69 (ethanol -H_2O, 1:2)	-	-	12.18
Xh	4-HO-3,5-(*tert*-C_4H_9)$_2C_6H_2$-	73 (A), 84 (F)	87-94 (benzene-hexane, 2:1)	-	-	9.74

Physical-chemical characteristics of synthesized derivatives of alkylthioacetic acids (Xa-h, XIa-g, XIIIa-f and XIVa-e) are presented in Tables 2.15-2.18. The structures of compounds are confirmed by the data of IR- and PMR-spectroscopies and also elementary analysis. In the IR-spectra of these compounds stretching vibrations of C–S bonds are characterized by absorption maximums in the region 1210-1170cm^{-1} [167].

Table 2.17. Constants, yield and data of elementary analysis of N-(alkylthioacetyl)hydrazones of carbonyl compounds $AlkSCH_2CONHN{=}CRR^1$

Compound	R	R^1	Yield, %	$T_{melt.}$, °C (solvent for crystallization)	S,mass %
XIIIa	*iso*-C_3H_7-	H-	84	94-101 (ethanol-H_2O, 2:1)	15.72
XIIIb	Ph-	H-	77	124-129 (ethanol -H_2O, 2:1)	13.52
XIIIc	4-HO-3,5-(*tert*-C_4H_9)$_2C_6H_2$-	H-	83	85-90*)	9.14
XIIId	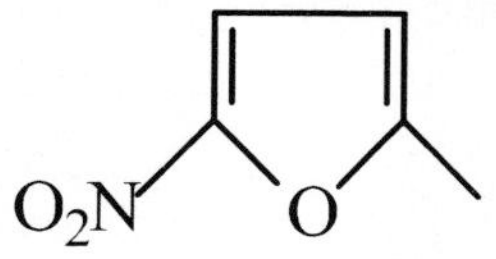	H-	87	145-149 (propanol-2)	11.05
XIIIe	CH_3-	CH_3-	78	oil*) (n_D^{20} 1,5534)	16.19
XIIIf	Ph-	CH_3-	81	73-79 (benzene-hexane, 2:1)	12.71

Note: *) Substances were purified by column chromatography on silicagel.

Table 2.18. Constants, yield and data of elementary analysis of N-(alkylthioacetyl)-N'-acylhydrazines $AlkSCH_2CONHNHCOR^2$

Compound	R^2	Yield, %	$T_{melt.}$, °C (solvent for crystallization)	S, mass %
XIVa	Ph-	78	144-150 (ethanol-H_2O, 3:1)	11.90
XIVb	4-$NO_2C_6H_4$-	83	167-174 (propanol-2)	11.08
XIVc	$PhCH_2$-	72	95-102 (ethanol-H_2O, 1:1)	11.63
XIVd	4-HO-3,5-(*tert*-C_4H_9)$_2C_6H_2SCH_2$-	76	130-136 (benzene-hexane, 3:1)	14.57
XIVe	$AlkSCH_2$-	80	72-82 (propanol-2-H_2O, 1:1)	21.05

In spectra of acids VIII there are the broad absorption band at 2760-2685cm^{-1} (ν HO), very strong band at 1720cm^{-1} (ν C=O) and absorption bands of average intensity at 1295cm^{-1} (ν C-O) and 920cm^{-1}, which are typical for α-substituted aliphatic acids [167]. In spectra of ethers Xa-h stretching vibrations of carbonyl groups are shifted to the high-frequency region and are revealed in the interval 1760-1740cm^{-1} and in spectra of chloroanhydrides IX – at 1775-1770cm^{-1}.

In IR-spectra of amides XI stretching vibrations of NH are presented by two absorption bands of average intensities at 3380 and 3195cm^{-1}, and in spectra of N-monosubstituted amides XIc-f there is one absorption band in the region 3230-3200cm^{-1}. Intensive maximums of absorption in the interval 1680-1645cm^{-1} should be assigned to stretching vibrations of C=O bond in amide group.

In IR-spectra of N-acylhydrazones XIIIa-f the intensive maximums of absorption in the region 1655-1630cm^{-1} correspond to stretching vibrations of C=N group [172]; absorption bands at 1690-1665cm^{-1} may be assigned to vibrations of N-acyl group of "amide I" type [167].

In PMR-spectra of synthesized compounds in the region of protons resonance of alkyl fragments there are several groups of signals differing in the values of chemical shifts. The signals of protons of methyl groups represent doublets and triplets at 0.89-1.12m.p.. Complex multiplets in the interval 1.18-1.38m.p. should be assigned to protons of methyl units, in the interval 1.68-1.89m.p. – to protons of CH-groups and signals shifted to the weak field (2.52-3.27m.p.) – to protons of CH_2- and CH-groups connected with sulfur atom. Singlet signals in the region 2.98-4.05m.p. characterize protons of acetyl fragments [167].

The individual compositions of mixtures of some synthesized derivatives of alkylthioacetic acids determined by the data of chromatography-mass spectrometry are presented in Table 2.19. These data also testify to close reactivity of homologous and isomers of mercaptans in the structure of their initial mixture in carried out chemical transformations.

Table 2.19. Individual composition of mixtures of synthesized alkylthioacetic acids derivatives (according to chromatography-mass spectrometry data)

Alk	The content of components in compounds mixture, %					
	Xb	Xf	XIg	XIIIb	XIIIe	XIVc
C_2H_5	3.0	3.5	2.9	3.5	3.3	3.0
n-C_3H_7	23.5	21.5	22.0	20.0	21.5	20.9
iso-C_3H_7	39.5	40.5	41.5	42.5	40.8	40.0
n-C_4H_9	5.5	5.5	4.5	5.5	5.0	5.2
iso-C_4H_9	22.5	23.5	23.4	22.0	22.2	24.4
n-C_5H_{11}	1.5	2.0	1.5	2.0	1.7	2.2
iso-C_5H_{11}	4.5	4.5	4.2	4.5	5.5	4.3

2.2.3. Synthesis of Five- and Six-Membered N-Containing Heterocycles on the Base of Derivatives of Alkylthioacetic Acids

Chloroanhydrides (IX), methyl ethers (Xa), hydrazides (XII) of alkylthioacetic acids amd also N-(alkylthioacetyl)-N'-acylhydrazides (XIVa-f) which synthesis and properties were described in Section 2.2.1 were used by us as synthons under production of various five- and six-membered nitrogen-containing heterocyclic compounds with alkylthiomethyl groups – Δ^2-imidazolines, Δ^2-oxazolines, benzimidazoles, benzoxazoles, 1,2,4-triazoles, 1,2,4-oxadiazoles, 1,3,4-oxadiazoles and *sym*-triazines.

It is known [173, 174] that condensation of ethylenediamine or N-substituted ethylenediamines with derivatives of carbonic acids (nitriles, esters, thioamides, iminoethers, amidines, etc.) leads to 2-substituted or 1,2-disubstituted Δ^2-imidazolines. In presented work for the reception of Δ^2-imidazolines of mentioned above type the condensation of methyl ethers Xa with ethylenediamine, N-butyl- and N-benzylethylenediamine was carried out (method A) by recommended in literature [175] boiling of ethers with the excess of corresponding diamine in waterless benzene in the presence of 30 mass. % of cationite KU-2x8.

We established that at molar ratio of reagents 1 : 5 and azeotropic distillation of water formed for 6-10 hours the Δ^2-imidazolines XVa-c were formed with the yield 51-59% (Table 2.20). Together with desired Δ^2-imidazolines the N-acyl derivatives of initial diamines – N,N'-di(alkylthioacetyl)ethylendiamines (XVI), N-butyl- (XVIIa) and N-benzyl-N'-(alkylthioacetyl)ethylenediamine (XVIIb) were isolated out of reaction mixtures with the yield 17-24% as by-products. The rise of ratio methyl ethers Xa : diamine up to 1 : 6–1 : 8 (moles) and duration of synthesis and also carrying out of the reaction in boiling toluene did not lead to significant decrease of yields of by-products XVI and XVIIa, b and increase of yield of Δ^2-imidazolines XVa-c.

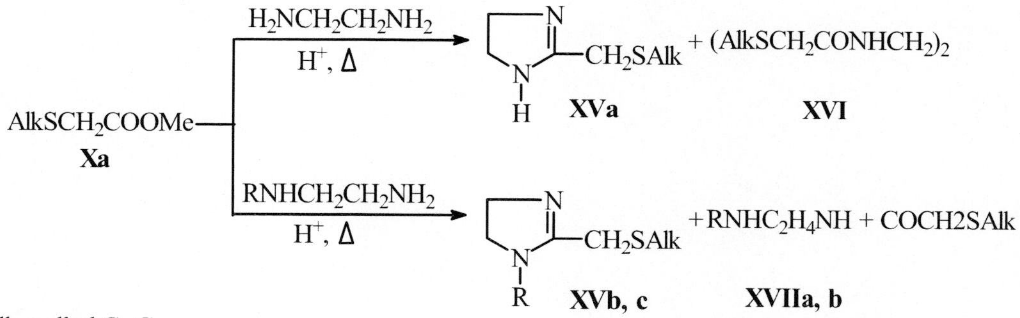

Alk = alkyl C_2-C_5;
R = n-C_4H_9 (XVb, XVIIa), $PhCH_2$ (XVc, XVIIb)

Table 2.20. Constants, yield and data of elementary analysis of 2-alkylthiomethyl derivatives of Δ^2-imidazolines, Δ^2-oxazolines, benzimidazoles and benzoxazoles

Compound	R	R[1]	Yield, %	$T_{melt.}$, °C (solvent for crystallization)	n_D^{20}	S, mass %
XVa	H	-	67	oil*)	1.5833	18.67
XVb	н-C_4H_9	-	56	oil*)	1.5774	14.69
XVc	$PhCH_2$	-	51	oil*)	1.5860	12.72
XVIIIa	H	-	61	oil*)	1.5752	18.92
XVIIIb	CH_3	-	65	oil*)	1.5794	18.08
XIXa	H	-	78	130-136 (ethanol-water, 1:1)	-	15.40
XIXb	Cl	-	70	154-161 (ethanol)	-	13.32
XXa	H	H	74	55-61 (petroleum ether)	-	14.82
XXb	H	NO_2	76	140-145 (DMPhA-water, 1:1)	-	11.88

Note: *) Substances were purified by column chromatography on silicagel.

2-Alkylthiomethyl-Δ^2-imidazolines XVa were also synthesized with the yield 65-67% by condensation of ethers Xa with ethylenediamines and dihydrochloride of ethylenediamine in ethyleneglycole in the presence of p-toluenesulfoacid at 130-135°C for 16 hours (method B

[175]). In this case the reaction leads to only one desired product – Δ^2-imidazoline XVa and is not accompanied by formation of by-product XVII.

Table 2.21. Constants, yield and data of elementary analysis of 3-R-5-alkylthiomethyl-1H-1,2,4-triazoles

Compound	R	Yield, % (preparative method)	$T_{melt.}$, °C (solvent for crystallization)	S, mass %
XXIIIa	n-C_5H_{11}-	66 (A), 78 (B)	oilo*) (n_D^{20} 1,5954)	13.15
XXIIIb	Ph-	72(A), 80 (B)	84-91 (dioxane-water, 1:1)	13.61
XXIIIc	$PhCH_2$-	68 (A), 75 (B)	52-58*)	12.09
XXIIId	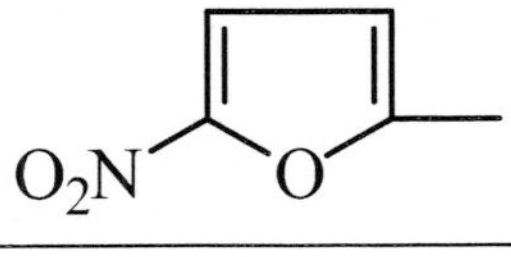	71 (A), 81 (B)	148-154 (ethanol-water, 5:1)	10.56
XXIIIe	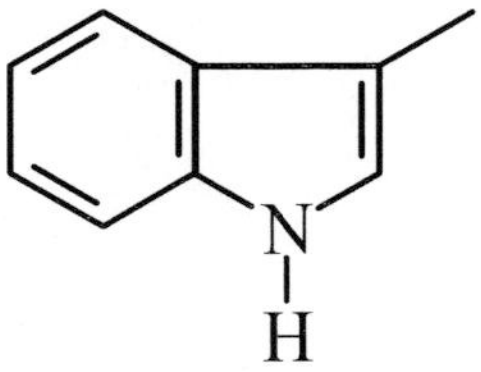	73 (A), 67 (B)	120-124 (ethanol-water, 2:1)	13.86
XXIIIf	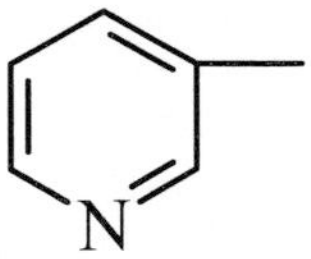	70 (A), 82 (B)	97-103 (benzene-hexane, 3:1)	9.81

Note: *) Substances were purified by column chromatography on silicagel.

2-alkylthiomethyl-5-R-Δ^2-oxazolines (XVIIIa, b) were obtained by condensation of methyl ethers Xa with monoethanolamine and 1-aminopropanol-2 (molar ratio 1 : 5) in the presence of cationite KU-2x8 in toluene (boiling for 8 hours) with the yield 60-65%. Analogously from ethers Xa and 1,2-diamino-4-R-benzenes and 2-amino-4-R-5-R'-phenoles (molar ratio 1 : 1.5) with the yield 70-78% correspondingly the 2-alkylthiomethyl-5-R-benzimidazole (XIXa, b) and 2-alkylthiomethyl-5-R-6-R'-benzoxazoles (XXa, b) were synthesized.

In accordance with [177] N^1-acylamiderazones of carbonic acids may serve as initial substances for synthesis of 3,5-disubstituted 1H-1,2,4-triazoles. In presented work we used as initial compounds N^1-(alkylthioacetyl)-amiderazones of aliphatic, aromatic and heterocyclic acids (XXIa-f). N^1-(alkylthioacetyl)amiderazones XXIa-f are formed with the yield 81-92% (Table 1 in the Appendix) under condensation of equimolar amounts of hydrazides of alkylthioacetic acids XII and ethyl iminoethers of caproic (XXIIa), benzoic (XXIIb), phenylacetic (XXIIc), 5-nitro-2-furancarbonic (XXIId), indol-3-carbonic (XXIIe) and nicotinic (XXIIf) acids in methanol (boiling for 0.5-1 hours).

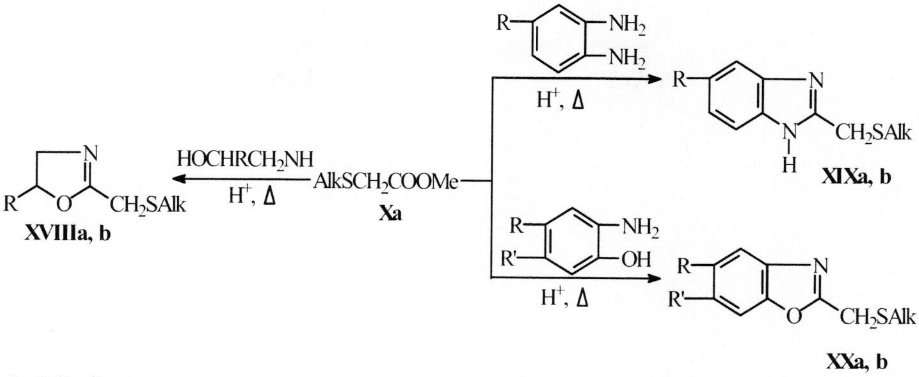

Alk = alkyl C_2-C_5;
R = H (XVIIIa, XIXa), CH_3 (XVIIIb), Cl (XIXb);
R = R^1 = H (XXa); R = H, R^1 = NO_2 (XXb)

Cyclodehydration of N^1-acylamiderazones XXIIIa-f with formation of 3-substituted 5-(alkylthiomethyl)-1H-1,2,4-triazoles (XXIIIa-f) proceeds when heating up to the temperature by 5-10°C higher than their melting temperature for 5-10 minutes (method A) or under the action of excess of thiolnylchloride in waterless ether or chloroform at 20-40°C for 2-3hours (method B).

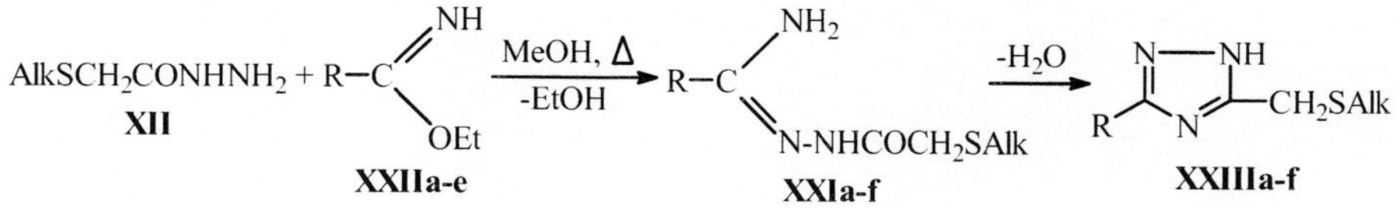

Alk = alkyl C_2-C_5;
R = n-C_5H_{11} (a), Ph (b), $PhCH_2$ (c), 5-nitrofuryl-2 (d), indolyl-3 (e), pyridile-3 (f)

One of the most commonly used method of reception of 1,2,4-oxadiazoles is cyclodehydration of O-acylamidoximes of carbonic acids [173]. This method was used for synthesis of 3-substituted 5-alkylthiomethyl-1,2,4-oxadiazoles (XXIVa-f). Acylation of amidoximes of acetic (XXVa), chloroacetic (XXVb), benzoic (XXVc), phenylacetic (XXVd), 5-nitro-2-furancarbonic (XXVe) and nicotinic (XXVf) acids (synthesis and properties are described in the Appendix) by chloroanhydrides of alkylthioacetic acids IX in pyridine (45-50°C, 2 hours) leads to O-(alkylthioacetyl)amidoximes XXVIa-f of corresponding carbonic acids with the yield 83-91%. When heating the last ones in vacuum (10-15 millimeters of mercury) at temperature by 3-5°C higher than their melting temperature the violent decomposition occurs with formation of coloured residue out of which 1,2,4-oxadiazoles XXIVa-f are isolated with the yield 57-65% (Table 2.22).

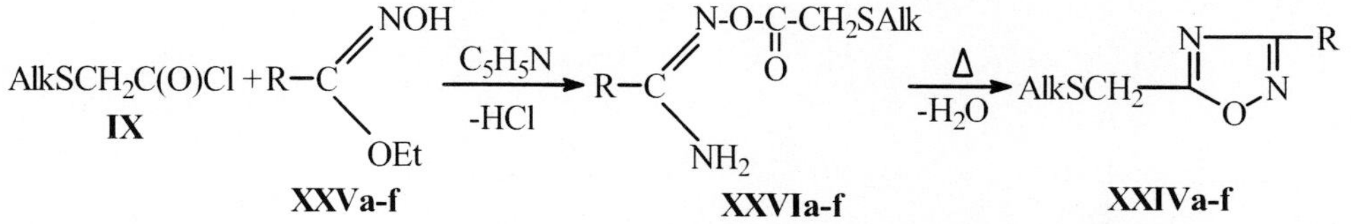

Alk = alkyl C_2-C_5; R = CH_3 (a), $ClCH_2$ (b), Ph (c), $PhCH_2$ (d), 5-nitrofuryl (e), pyridile-3 (f)

Table 2.22. Constants, yield and data of elementary analysis of 3-R-5-alkylthiomethyl-1,2,4-oxadiazoles

Compound	R	Yield, %	$T_{melt.}$, °C (solvent for crystallization)	n_D^{20}	S, mass %
XXIVa	CH_3-	59	oil*)	1.5694	18.29
XXIVb	$ClCH_2$-	60	oil*)	1.5781	14.81
XXIVc	Ph-	65	65-69 (petroleum ether)	-	13.81
XXIVd	$PhCH_2$-	57	oil*)	1.5839	12.03
XXIVe	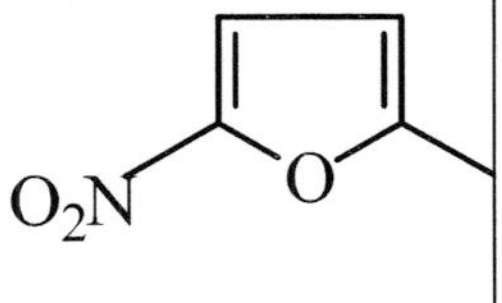	62	82-87 (ethanol-water, 1:1)	-	11.12
XXIVf	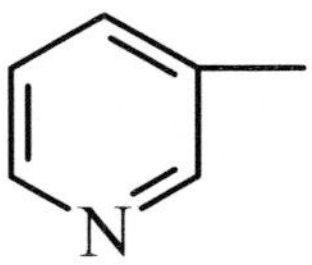	63	101-105 (ethanol-water, 1:1)	-	10.62

Note: *) Substances were purified by column chromatography on silicagel.

Cyclodehydration N-(alkylthioacetyl)-N'-acylhydrazines XIVa-f by the known method [178] under the action of phosphorus oxichloride leads to 5-substituted 2-alkylthiomethyl-1,3,4-oxadiazoles XXVIIa-f. During short-time heating of N,N'-diacylhydrazines XIVa-e with $POCl_3$ and T = 75-80°C the 1,3,4-oxadiazoles XXVIIa-e are formed with the yield 64-73% (Table 2.23).

Table 2.23. Constants, yield and data of elementary analysis of 2-alkylthiomethyl-5-R-1,3,4-oxadiazoles

Compound	R	Yield, %	$T_{melt.}$, °C (solvent for crystallization)	S, mass %
XXVIIa	Ph-	70	101-108 (heptane)	12.62
XXVIIb	4-$NO_2C_6H_4$-	73	148-154 (ethanol-water, 1:1)	11.55
XXVIIc	$PhCH_2$-	67	72-78 (heptane)	12.35
XXVIId	4-HO-3,5-(*tert*-$C_4H_9)_2C_6H_2SCH_2$-	66	oil*)	14.87
XXVIIe	$AlkSCH_2$-	64	oil*) (n_D^{20} 1,5881)	21.91

Note: *) Substances were purified by column chromatography on silicagel.

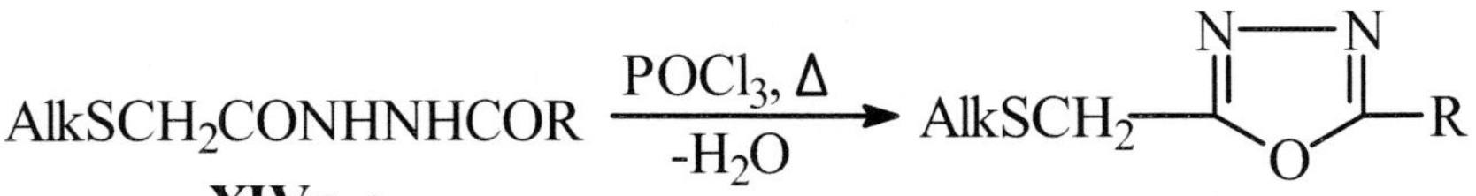

XIVa-e **XXVIIa-e**

Alk = alkyl C_2-C_5;

R = Ph (a), 4-$NO_2C_6H_4$ (b), $PhCH_2$ (c), 4-HO-3,5-(*tert*-$C_4H_9)_2C_6H_2SCH_2$ (d), $AlkSCH_2$ (e)

It is known [179, 180] that N-substituted 2,4-diamino-6-alkyl(aryl, heteryl)-*sym*-triazines (guanamines) are formed as a result of condensation of esters with biguanides. However in spite of wide application of various esters in this reaction there are no data in literature about participation of any esters of alkylthioacetic acids (individual or in mixtures) in condensation with bioguanidine until now. In connection with this we studied in details the condensation of mixture of methyl ethers Xa with hydrochlorides of N-mono- and N,N-disubstituted biguanides XXVIIIa-i and by the reaction:

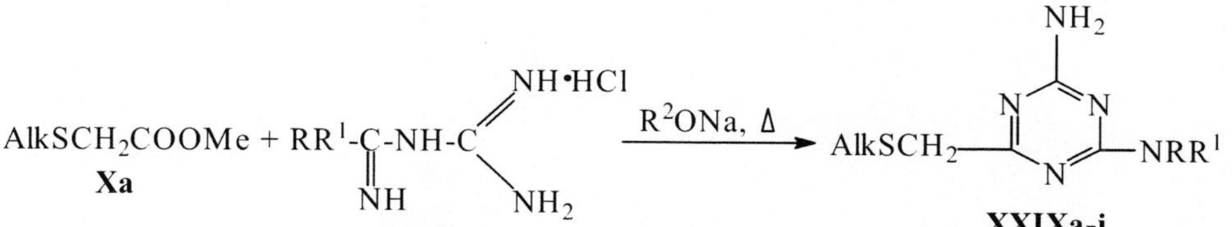

Xa **XXVIIIa-i** **XXIXa-i**

Alk = alkyl C_2-C_5;
R = CH_3, R^1 = H (a); R = R^1 = CH_3 (b); R = CH_3, R^1 = *n*-$C_{18}H_{37}$ (c); R = *cyclo*-C_6H_{11}, R^1 = H (d); R = Ph, R^1 = H (e); R = 4-$CH_3C_6H_4$, R^1 = H (f); R = R^1 = $PhCH_2$ (g); RR^1N = morpholino (h); RR^1N = piperidino (i); R^2 = CH_3, *iso*-C_3H_7

We showed that it was advisable to carry out the condensation of methyl ethers **Xa** with hydrochlorides of bioguanidines XXVIIIa-i in two stages. At the first stage free bases were isolated out of biguanides salts; toward this end the most suitable reagent is the solution of sodium alcoholate in corresponding alcohol – methanol or propanol-2. The second stage was carried out by boiling (10-18 hours) of equimole amounts of biguanide bases and methyl ether Xa in the same solvent in the presence of catalytic amounts of sodium alcoholate; at these conditions the desired 2-amino-4-(RR^1-amino)-6-alkylthiomethyl-*sym*-triazines XXIXa-i are formed with the yield 48-84% (Table 2.24).

We found that reaction time and yield of desired *sym*-triazines XXIXa-j depended on nature and reactivity of both initial methyl ethers Xa and N-substituted biguanides. The reagents are usually boiled in ethanol for 4-5 hours under reception of N-substituted 2,4-diamino-6-alkyl-*sym*-triazines on the base of ethers of aliphatic acids. It turned out that cyclocondensation of bioguanides with ethers of alkylthioacetic acids Xa should be carried out with boiling of reagents for 10-18 hours. We should note that replacing of methanol or propanol-2 by higher-boiling solvents (dioxane, butanol-1, DMFA) did not lead to noticeable increase of the yield of desired products. Obviously lower reactivity of methyl ethers of thioacetic acids Xa in comparison with ethers of aliphatic acids is caused by stronger +I-effect of alkylthiol group significantly decreasing ability of carbonyl group to connect bioguanides.

At the same time N,N-dimethyl- (XVIIIb), N,N-(3-oxapentamethylene)- (XVIIIh) and N,N-pentamethylenebiguanide (XVIIIi) react with ethers Xa more active then other biguanides. So, under heating of N,N-dimethylbiguanide XVIIIb and methyl ethers Xa in propanol-2 for 10 hours the yield of 2-amino-4-dimethylamino-6-alkylthiomethyl-*sym*-triazines XXIXb reaches 78%. At analogous conditions from N-methyl-N-octadecyl- (XXVIIIc) and N,N-dibenzylguanide (XXVIIIg) the corresponding *sym*-triazines XVIIIc, g are formed only with the yield 48-54%. Probably such difference in reactivity is connected with the rise of spatial difficulties created by volumetric substituents in initial biguanides.

Table 2.24. Constants, yield and data of elementary analysis of 2-amino-4-(RR'-amino)-6-alkylthiomethyl-*sym*-triasines

Compound	RR'N	Yield (preparative method*)), %	$T_{melt.}$, °C (solvent for crystallization)	S, mass %
XXIXa	CH_3NH-	66 (A)	132-138 (butanol-1)	14.11
XXIXb	$(CH_3)_2N$-	78 (A) 86 (B)	110-114 (ethanol-water, 3:1)	14.02
XXIXc	*n*-$C_{18}H_{37}(CH_3)N$-	54 (A)	oil**)	6.48
XXIXd	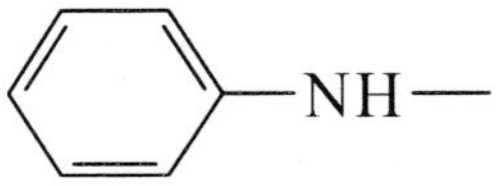	70 (A)	150-156 (propanol-2)	11.38
XXIXe	PhNH-	72 (A) 77 (B)	147-152 (dioxane-water, 2:1)	11.81
XXIXf	4-$CH_3C_6H_4NH$-	54 (A) 82 (B)	165-169 (dioxane-water, 2:1)	10.50
XXIXg	$PhCH_2N(-)CH_2OH$	48 (A)	101-105 (ethanol)	9.78
XXIXh	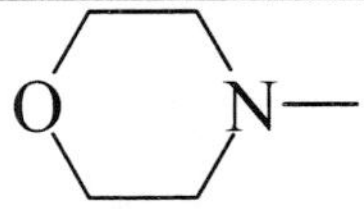	84 (A) 88 (B)	76-81 (acetonitrile)	12.15
XXIXi	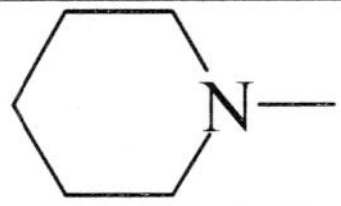	80 (A) 83 (B)	94-99 (ethanol-water, 1:1)	11.22

Notes:
*) method A – condensation of N-substituted biguanides with methyl ethers of alkylthioacetoc acids; method B – interaction of N-substituted 2,4-diamino-6-chloromethyl-*sym*-triazines with mercaptans C_2–C_5 fraction;
**) Substances were purified by column chromatography on silicagel.

The necessary condition of active cyclocondensation is the presence of basic catalyst (sodium alcoholate) in reaction mixture on the second stage of process. For example, when we used biguanide XVIIIh the absence of base even after heating for 72 hours led to essential content of unreacted methyl ether Xa and yield of desired 2-amino-4-morfolino-6-alkylthiomethyl-*sym*-triazines XXIXh did not exceed 25-28%.

Physical-chemical characteristics of synthesized heterocyclic compounds are presented in Tables 2.20–2.24. The structures of obtained heterocycles are confirmed by data of IR- and PMR-spectroscopies, and also by elementary analysis. In IR-spectra of all compounds stretching vibrations of C–S bonds are characterized by absorption bands of variable intensity in the region 1205–1155cm^{-1} [167].

In IR-spectra of Δ^2-imidazolines XVa-c there are strong absorption bands in the region 1625-1605cm^{-1} characteristic for stretching vibrations of C=N group in cyclic systems without internal conjugation [181, 182]. Position of this band is not changed under salt formation that distinguishes Δ^2-imidazolines from other compounds with C=N group which transformation into salts is accompanied by shift of analogous band to high-frequency region [183]. Stretching and deformation vibrations of methylene groups of imidazoline cycle are presented by absorption bands of variable intensity in region 2915-2880, 1460-1445cm^{-1}

("cutter" vibration of CH_2), 1250-1240 and 1015-1005cm^{-1} correspondingly [181,182]. In the spectra of 2-substituted Δ^2-imidazolines XVa stretching vibrations of NH group of imidazoline cycle are revealed as broad band at 3370-3355cm^{-1} [182].

In spectra of compounds XVIIIa, b absorption bands of C=N group are revealed at 1650-1640cm^{-1} that is typical for double bond in Δ^2-oxazolines [184].

In IR-spectra of N^1-acylhydrazones XXIIa-f there are intensive maximums of absorption in the regions 3350-3340 and 3150-3135cm^{-1} characteristic for stretching vibrations of NH_2– and NH–groups of amiderazone fragment [185]. Absorption bands in the regions 1675-1655 and 1635-1620cm^{-1} should be assigned to stretching vibrations of C=O and C=N groups correspondingly.

In spectra of 1,2,4-triazoles XXIIIa-e there is a number of absorption bands caused by vibrations of triazole cycle: 1475-1460, 1505-1500, 1120-1110cm^{-1} (stretching vibrations of cycle), 1050-950cm^{-1} (deformation vibrations of cycle) and 1290-1275cm^{-1} (ν C-N in 1,2,4-triazoles [183,186]).

In IR-spectra of 1,2,4-oxadiazoles XXIVa-f absorption bands of C=N group are revealed at 1635-1620cm^{-1} that is characteristic for this bond in such oxadiazoles [183]. Absorption bands of average intensities in regions 1590-1570, 1465-1440 and 1400-1370cm^{-1} are conditioned by stretching and in region 910-890cm^{-1} by planar deformation vibrations of cycle of 1,2,4-oxadiazole [183]. Absorption maximums at 1415-1410, 1380-1365 and 1080-1065cm^{-1} are characteristic for stretching vibrations of bond =N–O–.

In spectra of 1,3,4-oxadiazoles XXVIIa-e there are intensive absorption maximums in the intervals 1615-1510 and 1495-1470cm^{-1} characteristic for stretching vibrations of oxadiazole cycle [183, 187]. The presence of the last one is confirmed also by absorption bands at 1250-1225 and 1050-1020cm^{-1} relating to stretching vibrations of =C–O–C= fragment in 1,3,4-oxadiazoles [188] and absorption maximums at 950cm^{-1} connected with "breathing" vibrations of oxadiazole cycle [183].

In IR-spectra of diamino-*sym*-triazines XXIXa-i there are absorption maximums of variable intensity characteristic for stretching (1570-1555, 1530-1520, 1420-1415cm^{-1}), "breathing" (1120-1105, 1010-995cm^{-1}), extra-planar (820-805cm^{-1}) and planar (715-695cm^{-1}) deformation vibrations of cycle of *sym*-triazine [180, 183]. In the region of stretching vibrations of NH in spectra of *sym*-triazines XXIXb, c, f, g-i there are two broad absorption bands in the intervals 3470-3375 and 3190-3120cm^{-1} (ν_{as} and ν_s of NH in NH_2-groups) [180]. In spectra of compounds XXIXa, d-f stretching vibrations of NH_2– and NH-groups are imposed and observed as broad absorption band in the region 3410-3180cm^{-1} with feebly marked separation of maximums.

In PMR-spectra of Δ^2-imidazolines XVa-c signals of 4 protons of imidazolone cycle are revealed as non-symmetric multiplet at 4.08–4.42m.p. [175]. In spectra of Δ^2-oxazolines XVIIIa, b the signals of protons of cycle's methylene groups are presented by multiplet in the interval 4.27–4.48m.p.

In PMR-spectra of *sym*-triazines XXIXa-i signals of protons of primary amino-groups are presented as broadened singlets with intensity equal to 2 proton units in the interval 5.76-6.36m.p.

For some of heterocyclic substances which synthesis was described in presented Section the contents of individual components in them confirming close reactivities of initial mixtures of reagents were determined by chromatography-mass spectrometry (Table 2.25).

Table 2.25. Individual composition of mixtures of synthesized nitrogen-containing heterocyclic compounds (according to the chromatography-mass spectrometry data)

Alk	Contents of components in mixture of compounds, %				
	XVc	XVIIIc	XXIIIb	XXVIIa	XXIXh
C_2H_5	3.5	3.5	2.8	3.3	3.5
n-C_3H_7	20.8	19.5	20.3	19.9	20.0
iso-C_3H_7	40.2	41.5	40.0	41.4	41.5
n-C_4H_9	5.5	5.0	5.1	5.0	4.0
iso-C_4H_9	24.0	23.5	25.1	23.7	25.0
n-C_5H_{11}	1.5	2.0	1.9	1.8	1.5
iso-C_5H_{11}	4.5	5.0	4.8	4.9	4.5

2.3. Synthesis of Alkylthio- and Alkylthiomethyl Derivatives of Five- and Six-Membered Heterocyclic Compounds by Alkylation Reactions

Derivatives of five- and six-membered heterocyclic compounds containing alkylthio- and alkylthiomethyl groups traditionally attract researchers' attention due to their high and versatile biological activity and also us substances with a wide spectrum of application properties.

The reception of a number of nitrogen-containing heterocycles of such type with two and three heteroatoms (Δ^2-imidazolines, Δ^2-oxazolines, benzimidazoles, benzoxazolesm 1,2,4-triazoles, 1,2,4- and 1,3,4-oxadiazoles) by cyclocondensation reactions with participation of functional derivatives of alkylthioacetic acids (chloroanhydrides, esters, hydrazides, N,N'-diacylhydrazines) was described in Section 2.2.2. In given Section we present synthesis of alkylthiomethylheterocycles by reaction of alkylation of mixture of natural mercaptans C_2–C_5 by reactive chloromethyl derivatives of various heterocyclic compounds. As alkylating agents we chose first of all available five-membered chloromethylheterocycles: tetrahydrofurfurylchloride, 2-chloromethylthiophene, 1-R-chloromethyl-Δ^2-imidazolines, 1-R-2-chloromethyl-Δ^2-imidazolines, 2-chloromethylbenzimidazol, 2-chloromethyl-6-R-benzoxazoles, 2-R-4-chloromethylthiazoles, 1-chloromethylbenztriazole, 2-chloromethyl-5-R-1,3,4-oxadiazoles and 3-chloromethyl-5-R-1,2,4-oxadiazoles.

$$\text{AlkSH} + \text{HetCH}_2\text{Cl} \xrightarrow[\text{-NaCl}]{\text{NaOH}} \text{HetCH}_2\text{SAlk}$$

XVa, d; XIXa; XXa, b;
XXVIIa, b, g; XXXa-g

Alk = alkyl C_2-C_5;

Het = -Δ^2-imidazolyl (XVa), 1-methyl-Δ^2-imidazolyl-2 (XVd), benzimidazolyl-2 (XIXa), benzoxazolyl-2 (XXa), 6-nitrobenzoxazolyl-2 (XXb), 5-phenyl-1,3,4-oxadiazolyl-2 (XXVIIa), 5-(4-nitrophenyl)-1,3,4-oxadiazolyl-2 (XXVIIb), tetrahydrofuryl-2 (XXXa), thienyl-2 (XXXb), 2-phenylthiazolyl-4 (XXXc), 2-(pyridile-3)thiazolyl-4 (XXXd), benztriazolyl-1 (XXXe), 5-phenyl-1,2,4-oxadiazolyl-3 (XXXf), 5-(pyridile-3)-1,2,4-oxadiazolyl-3 (XXXg).

Alkylation of mercaptans C_2–C_5 was carried out in methanol or propanol-2 at 45-65°C and water solution of alakli was used as acceptor of HCl. Reaction time is determined first of all by the nature of initial alkylating reagent. For example under the use of tetrahydrofurylchloride and 1-Cl-mehtylbenztriazole after boiling in methnol for 10-12 hours the desired alkyl-(tetrahydrofurfuryl)sulfides XXXa and 1-alkylthiomethylbenztriazoles XXXe were obtained with the yield 64-68%. In the rest cases corresponding alkyl-(heterylmethyl)sulfides (XVa, d; XIXa; XXa, b; XXVIIa, b; XXXb-d, f, g) were obtained with high yield (Table 2.26) under heating in methanol or propanol-2 at 45-50°C for 3-5 hours.

Table 2.26. Constants, yield and data of elementary analysis of alkylthiomethyl derivatives of five-membered heterocyclic compounds $HetCH_2Salk$

Compound	Het	Yield, %	$T_{boil.}$, °C (millimeters of mercury) or $T_{melt.}$, °C (solvent for crystallization)	n_D^{20}	S, mass %
XVa	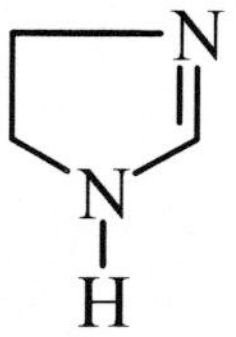	72	oil*)	1.5830	18.94
XVd	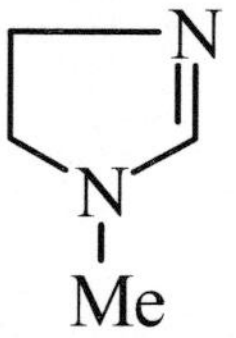	74	111-135 (2)	1.5778	19.37
XIXa	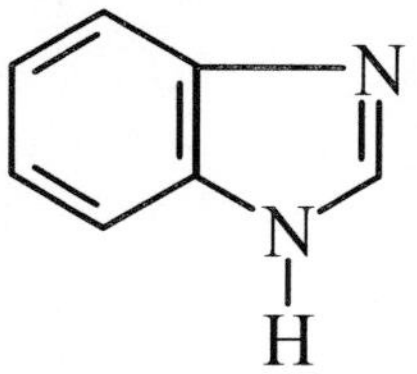	82	131-136 (ethanol-water, 1:1)	-	14.84
XXa	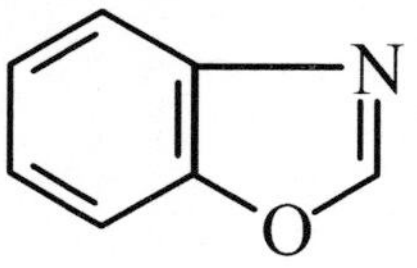	78	56-60 (petroleum-ether)	-	15.24
XXb	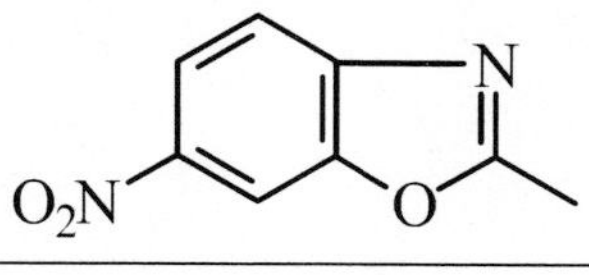	70	142-147 (DMFA-water, 1:1)	-	12.55
XXVIIa	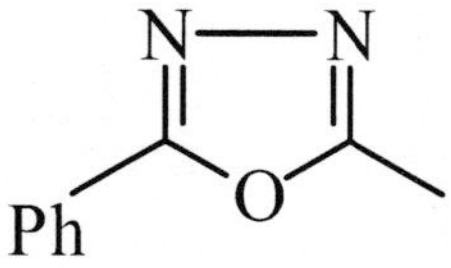	75	102-107 (heptane)	-	13.57

Table 2.26. (Continued)

Compound	Het	Yield, %	$T_{boil.}$, °C (millimeters of mercury) or $T_{melt.}$, °C (solvent for crystallization)	n_D^{20}	S, mass %
XXVIIb	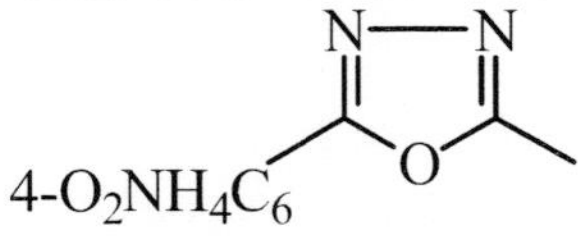	70	150-155 (ethanol-water, 1:1)	-	11.24
XXXa	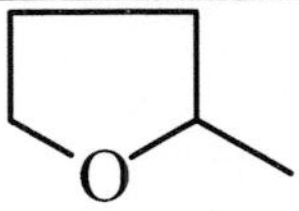	68	88-113 (4)	1.5715	19.31
XXXb	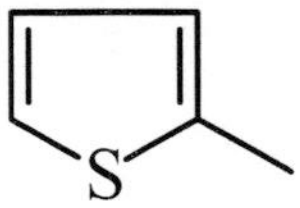	65	98-124 (3)	1.5868	35.67
XXXc	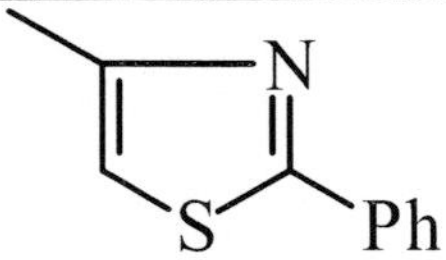	76	122-127 (dioxane-water, 2:1)	-	24.57
XXXd	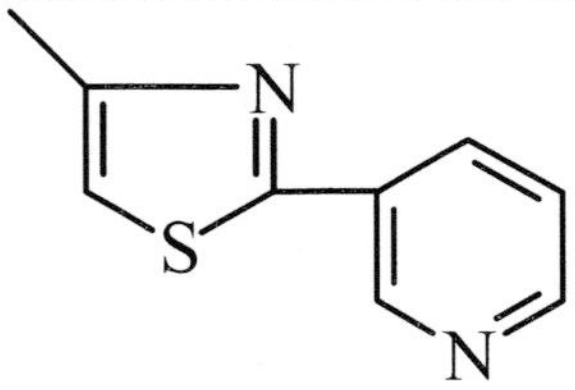	70	164-169 (DMFA-water, 1:2)	-	24.16
XXXe	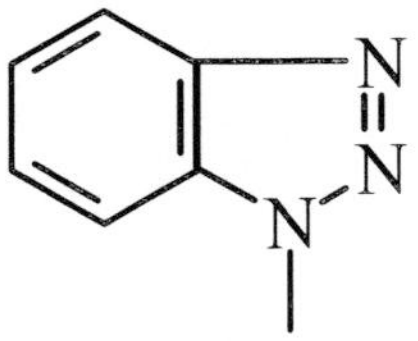	64	84-88 (benzene-hexane, 2:1)	-	14.83
XXXf	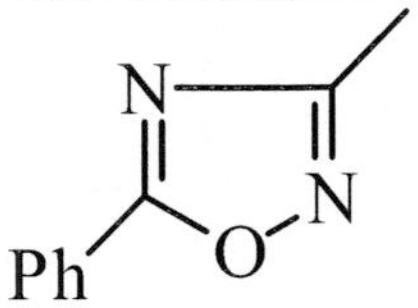	73	oil*)	1.5904	12.85
XXXg	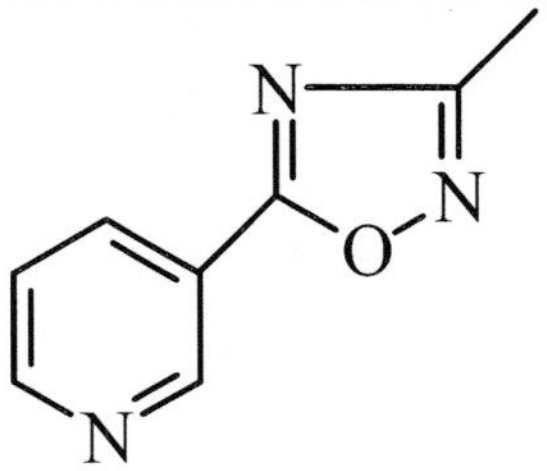	66	115-120 (dioxane-water, 1:1)	-	13.21

Note: *) Substances were purified by column chromatography on silicagel.

It is known [189, 190] that in 2,4-diamino-6-chloromethyl-*sym*-triazines the chlorine atom is easily substituted under the action of amines and mercaptans. We used this method as alternative one under reception of N-substituted 2,4-diamino-6-alkylthiomethyl-*sym*-triazines XXIXb, e, h, i, j by interaction of N-substituted 2,4-diamino-6-chloromethyl-*sym*-triazines XXXIa-e with mercaptans C_2–C_5 mixture. The initial chloromethyl-*sym*-triazines XXXIa-e were synthesized by presented in works [188, 189] methods of condensation of N-substituted bioguanides XXVIIIb, e, f, h, i with methyl ether of chloroacetic acid in methanol in the presence of sodium methylate with the yield 60-77%.

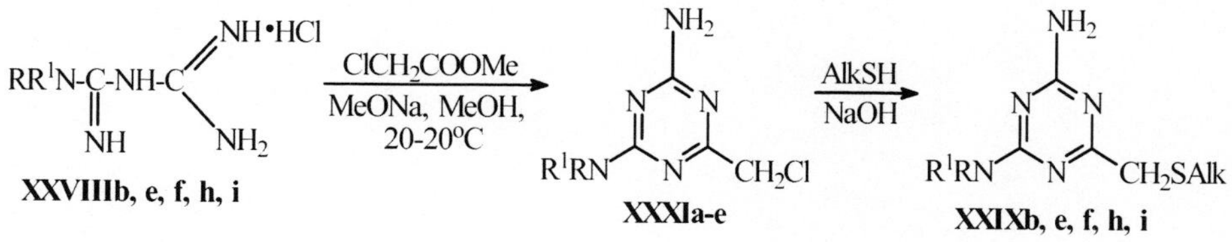

Alk = alkyl C_2–C_5;
R = R[1] = CH_3 (XXVIIIb, XXIXb, XXXIa);
R = Ph, R[1] = H (XXVIIIe, XXIXe, XXXIb);
R = 4-$CH_3C_6H_4$, R[1] = H (XXVIIIf, XXIXe, XXXIc); R[1]RN = morpholino (XXVIIIh, XXIXh, XXXId); pyperidino (XXVIIIi, XXIXi, XXXIi)
Alk = alkyl C_2-C_5.

It was established experimentally that interaction of chloromethyl-*sym*-triazines XXXIa-e with mercaptans C_2–C_5 mixture should be carried out with boiling (4-5 hours) in aqueous alcohol in the presence of some excess of alakli (1 : 1 : 1.05). At these cases the yield of desired 2-amino-4-(R[1]R-amino)-6-alkylthiomethyl-*sym*-triazines XXIXb, e, f, h, i is 77-88% (Table 2.24, method B).

Moreover, the same mercaptans C_2–C_5 were alkylated by C1-derivatives of N-heterocycles containing mobile chlorine atom [172]: 2-C1-benzimidazole, 2-C1-benzoxazole, 2-C1-benzthiazole, etc. (Table 2.27).

We showed that in these cases it is necessary to carry out alkylation by reaction presented below at more severe conditions: at long-continued (10–14 hours) heating of reagents in aqueous dioxane or aqueous DMFA in the presence of alakli excess (molar ratio 1 : 1 : 1.5). At these conditions the yield of desired alkylheterylsulfides XXXIIa-e is 56-65% (Table 2.27).

$$\text{AlkSH} + \text{HetCl} \xrightarrow[\Delta]{\text{NaOH}} \text{Het-S-Alk}$$

XXXIIa-e

Alk = alkyl C_2-C_5;
Het = benzimidazolyl-2 (a), benzoxazolyl-2 (b), benzthiazolyl-2 (c), 4-phenylthiazolyl-2 (d), 3,5-dichloropyridile-2 (e).

The structures of synthesized hetero-cyclic sulfides are confirmed by data of elementary analysis, IR- and PMR spectroscopies. In the IR spectra of these compounds together with absorption bands characteristic for corresponding heterocyclic system there are absorption maximums of variable intensity in the region 1210–1160cm^{-1}, relating to stretching vibrations of bonds C–S [167]. In PMR-spectra alkyl(heterylmethyl)sulfides the resonance signals in the region 3.32-3.56m.p. characterize protons of methylene groups directly connected with heterocyclic ring.

Table 2.27. Constants, yield and data of elementary analysis of alkylheterylsulfides Het-S-Alk

Compound	Het	Yield, %	$T_{boil.}$, °C (millimeters of mercury) or $T_{melt.}$, °C (solvent for crystallization)	n_D^{20}	S, mass %
XXXIIa	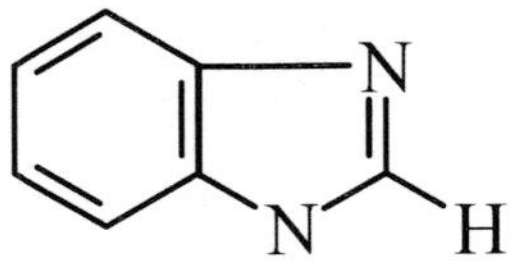	63	110-114 (ethanol-water, 2:1)	-	16.32
XXXIIb	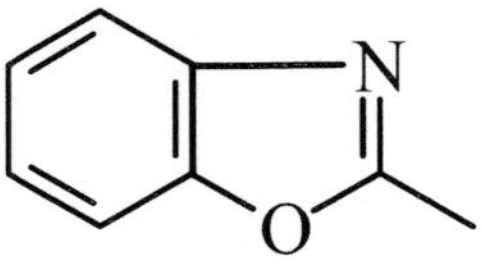	58	116-136 (3)	1.5752	15.47
XXXIIc	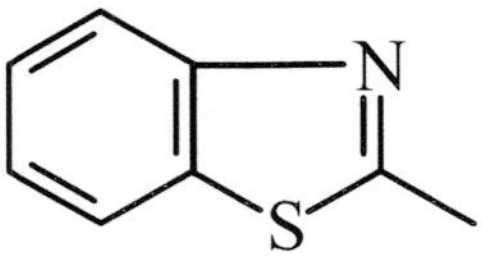	65	153-159 (3)	1.5835	29.85
XXXIId	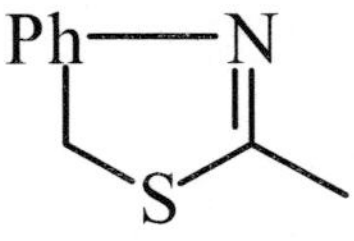	60	124-130 (ethanol-water, 1:1)	-	27.67
XXXIIe	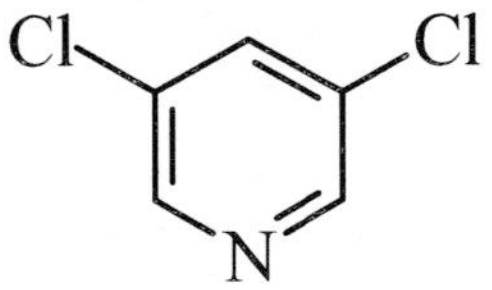	56	118-140 (2)	1.5934	15.31

2.4. Synthesis of 2-Alkylthio-4,6-Disubstituted and 2,4-Dialkylthio-6-Substituted Sym-Triazines

Theoretical and applied science of synthesis of various derivatives of *sym*-trazine possessing high chemical and thermal stability and correspondingly the wide spectrum of practical use are successfully developed at present [7, 48, 49, 142, 143, 191, 192]. High thermal stability of *sym*-triazine cycle is conditioned by high resonance energy (345.67 kJ/mole) due to its aromatic structure and presence of nonbonding electrons at nitrogen atom [193].

Among compounds of *sym*-triazine type the sulfur-containing derivatives (mercaptans, sulfides, sulfones, thiouronium salts, derivatives of dithioaminoformic acid, etc.) in which sulfur atoms are directly bonded with heterocyclic ring represent special interest. They may serve as high effective antioxidants and thermal stabilizers of polymer materials, rubbers and lubricating oils, vulcanizing agents for natural and synthetic rubbers, modifiers of properties of various polymers, etc. [142, 193]. The examples of industrial products are Irganox-565 – 2,4-bis-(octylthio)-6-(4-hydroxy-3,5-di-*tert*-butylphenoxy)-*sym*-triazine, Irganox 258 – 2-octylthio-4,6-bis-(4-hydroxy-3,5-di-*tert*-butylphenoxy)-*sym*-triazine and Antioxidant-912 – 2,4-bis-(2-octylthioethylthio)-6-(4-hydroxy-3,5-di-*tert*-butylphenoxy)-*sym*-triazine.

It is known [49] that derivatives of *sym*-triazine possess also the wide spectrum of pharmacologic activity, may serve as effective herbicides, fungicides and stimulators of plants growth [48, 191]. According to numerous literature data introduction into molecules of biologically active substances of alkylthiol group in some cases leads to significant modification of their biological activity. From this point of view the group of herbicides of novel generation is of great interest – 2-methylthio-4-isopropylamino-6-ethylamino- (ametrine), 2-methylthio-4,6-di(isopropylamino)- (prometrine), 2-methylthio-4-*tert*-butylamino-6-ethylamino- (terbutrine), 2-methylthio-4-azido-6-*tert*-butylamino-*sym*-triazine (aziprotrane) and some others 2-methylthio-4,6-disubstituted *sym*-triazines [48, 191, 195]. Preparations of this group were synthesized by substitution of chlorine atoms in corresponding chlorine derivatives of *sym*-triazine being known herbicides by methylthiogroups; they are characterized by high activity, low rates of application and fast inactivation in biosubstrates.

In connection with all said above the development of simple and convenient methods of synthesis of sulfides of *sym*-triazine type in which atom S is directly bonded with heterocycle on the base of mercaptans C_2–C_5 fraction and accessible chloro-derivatives of triazine type is perspective. That is why firstly we studied the interaction of mentioned mercaptans mixture and 2,4,6-trichloro-*sym*-triazine (cyanurchloride, CCh) with substitution of one or two chlorine atoms by alkylthiogroups:

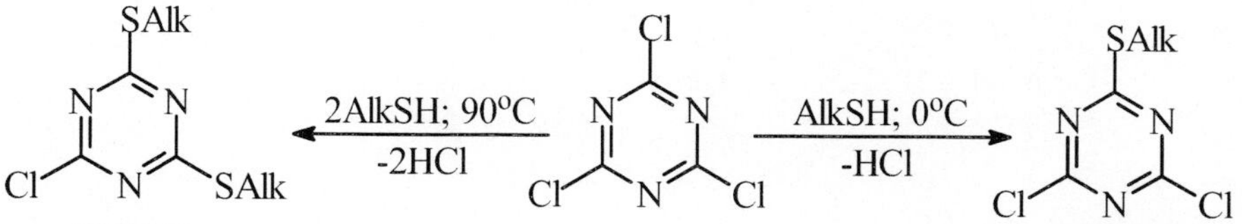

XXXIV **XXXIII**

Alk = alkyl C_2-C_5.

High mobility of chlorine atoms in cyanurchloride is explained by low electron density at carbon atoms of triazine cycle [196]. Substitution of chlorine atoms in cyanurchloride by electrono-donor groups (amino-, alkoxy-, alkylthio-, etc.) increases electron density at carbon atom and consequently decreases activity of non-reacted chlorine atoms. That is why substitution of chlorine atoms in cyanurchloride by amino- or alkoxygroup occurs by stages [193] as a consequence of some equalization of electron density in *sym*-triazine cycle. This increases strength of C–Cl bonding but decreases reactivity of chlorine atoms.

However under substitution of chlorine atoms in cyanurchloride by alkylthiogroups the rest chlorine atoms are deactivated to lesser degree than under introduction of amino- or alkoxygroup [7, 193]. Consequently under reception of 2,4-dichloro-6-alkyl(aryl)thio-*sym*-triazines also disubstituted derivatives may be formed in noticeable amounts even at relatively low temperature (-5–0°C) of reaction. Two methods are recommended in literature for synthesis of 2,4-dichloro-*sym*-triazines of such type:

- Interaction of cyanurchloride with mercaptans in acetone at 0°C with accepting of evolving HCl by collidine (2,4,6-trimethylpyridine) or lutidine (dimethylpyridine) taen in molar ratio 1 : 1 : 1 [197, 198], and
- Interaction of cyanurchloride with mercaptans at 0°C in toluene or chloroform in the presence of equivalent amount of water alkaline solution [198].

In our work for synthesis of 2-alkylthio-4,6-dichloro-*sym*-triazines XXXIII the second method was used. We established that at conditions mentioned above the desired dichloro-*sym*-triazines were obtained with the yield 60-64%, and as side-products (in accordance with chromatography-mass spectrometry data) 2,4-dialkylthio-6-chloro-*sym*-triazines XXXIV were formed in noticeable amount. The last ones were synthesized with the yield 58-62% by interaction of cyanurchlorid and mercaptans C_2–C_5 mixture (molar ratio 1 : 2) in aqueous dioxane at 30-35°C in the presence of NaOH.

The obtained 2-alkylthio-4,6-dichloro- XXXIII and 2,4-doalkylthio-6-chloro-*sym*-triazines XXXIV serve as convenient and reactive semi-products in synthesis of N-substituted 2,4-diamino-6-alkylthio-, 2-amino-4,6-dialkylthio-, 2-alkylthio-4,6-diaryloxy, 2,4-dialkylthio-6-aryloxy- and 2-alkylthio-4,6-diaryl(heteryl)thio-*sym*-triazines with alkyl, aryl and heteryl substituents including also the fragments of known antioxidants – spatially-hindered phenol, 2,2,6,6-tetramethylpyperidine, 4-aminodiphenylamine, 4-hydroxydiphenylamine (HODPA) and 2-mercaptobenzthiazole (MBT).

N-substituted 2,4-diamino-6-alkylthio-*sym*-triazines (XXXVa-m; N-substituted esters of thioameline) (method A) were synthesized with high yield as a result of interaction of 2,4-dichloro-*sym*-triazines XXXIII with primary and secondary amines of aliphatic, aromatic and heterocyclic type (molar ratio 1 : 2) in dioxane or DMFA at 100°C (Table 2.28). In all cases water solution of alakli was used as acceptor of HCl.

N-substituted 2,4-diamino-*sym*-triazines XXXVa, c, d, g, h, j-l were also synthesized by alternative method – by interaction of available N-substituted 2,4-amino-6-chloro-*sym*-triazines XXXVIa-i and mercaptans C_2–C_5 mixture in the presence of equivalent amount of alakli in aqueous dioxane or aqueous DMFA at 100–120°C during 8-12 hours (method B). N-substituted 2-amino-4,6-dialkylthio-*sym*-triazines (XXXVIIa-j, substituted esters of thioamelide) were synthesized with high yield (Table 2.29) analogously by two methods on the base of 2,4-dialkylthio-6-chloro-*sym*-triazones XXXIV and corresponding primary and secondary amines taken in molar ratio 1 : 1 (method A) or on the base of N-substituted 2-amino-4,6-dichloro-*sym*-triazines XXXVIIIa-f and mercaptans C_2–C_5 (1 : 2 moles) (method B).

Table 2.28. Constants, yield and data of elementary analysis of N-substituted 2,4-diamino-6-alkylthio-*sym*-triazines Salk

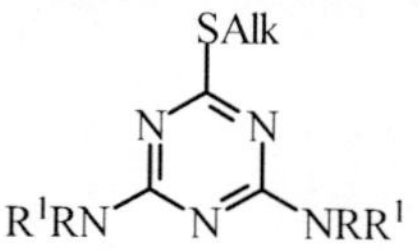

No.	R^1RN	Yield, % (preparative method)	$T_{melt.}$, °C (solvent for crystallization)	S, mass %
XXXVa	$(CH_3)_2N$-	70(A), 77(B)	98-105 (dioxane-water, 1:1)	13.13
XXXVb	*n*-$C_8H_{17}NH$-	72(A)	124-129 (ethanol-water, 3:1)	7.87
XXXVc	*n*-$C_{18}H_{37}(Me)N$-	66(A), 75(B)	54-60 (acetone)	7.15
XXXVd	$Ar^{*)}CH_2NH$-	70(A), 75(B)	164-168 (dioxane-water, 5:1)	5.07
No.	R^1RN	Yield, % (preparative method)	$T_{melt.}$, °C (solvent for crystallization)	S, mass %
XXXVe	4-$CH_3C_6H_4NH$-	64(A)	154-159 (ethanol)	8.31
XXXVf	4-ClC_6H_4NH-	71(A)	140-146 (DMFA-water, 2:1)	7.18
XXXVg	ArNH-*)	63(A), 67(B)	155-160 (toluene-heptane, 4:1)	5.04
XXXVh	4-$PhOC_6H_4NH$-	69(A), 64(B)	108-116 (propanol-2)	5.57
XXXVi	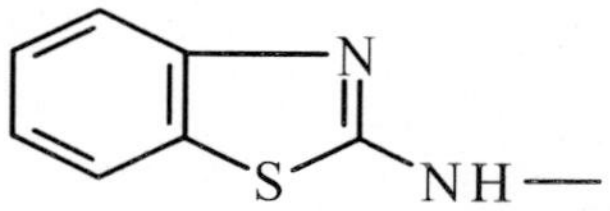	58(A)	135-140 (ethanol-water, 3:1)	20.75
XXXVj	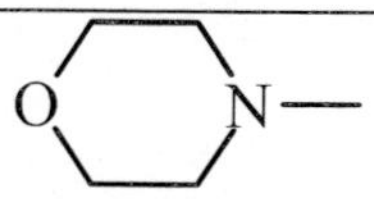	66(A), 70(B)	oil**) (n_D^{20} 1,5527)	10.14
XXXVk	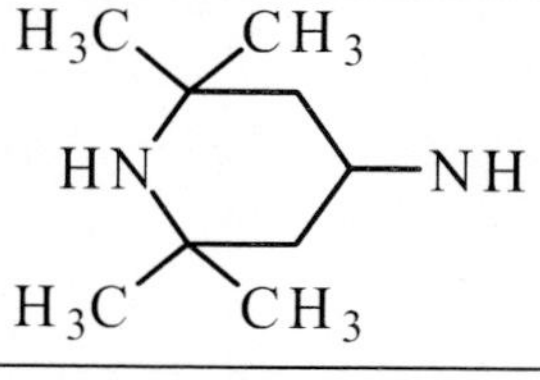	74(A), 82(B)	134-141 (dioxane-water, 2:1)	6.92
XXXVl	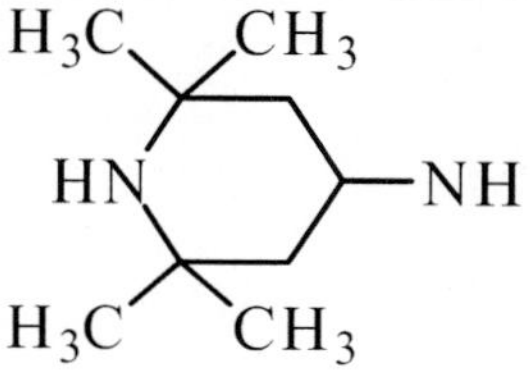 -$CH_2Ar^{*)}$	70(A), 67(B)	78-85 (petroleum-ether)	4.04
XXXVm	$(CH_3)_2N$	68(A), 75(B)	94-99 (dioxane-water, 1:1)	10.24

Notes: *) Ar = 4-HO-3,5-(*tert*-Bu)$_2C_6H_2$;

**) Substances were purified by column chromatography on silicagel.

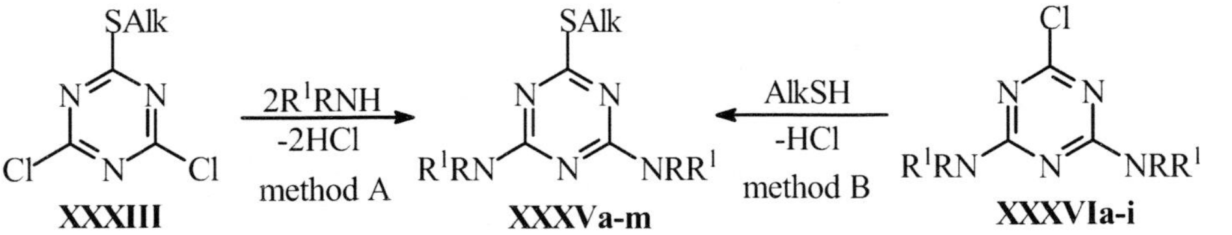

Alk = alkyl C_2-C_5;

R^1RN = $(CH_3)_2N$ (XXXVa, XXXVIa), *n*-$C_8H_{17}NH$ (XXXVb), *n*-$C_{18}H_{37}NCH_3$ (XXXVc, XXXVIb), Ar*)CH_2NH (XXXVd, XXXVIc), 4-$CH_3C_6H_4NH$ (XXXVe), 4-ClC_6H_4NH (XXXVf), Ar*)NH (XXXVg, XXXVId), 4-$PhOC_6H_4NH$ (XXXVh, XXXVIe), benzthiazolyl-2-amino (XXXVi), morpholino (XXXVj, XXXVIf), 2,2,6,6-$(CH_3)_4$-pyperidile-4-amino (XXXVk, XXXVIg);

R = CH_2Ar*), R^1 = 2,2,6,6-$(CH_3)_4$- pyperidile-4-amino- (XXXVl, XXXVIh), R^1R = $(C_2H_5)_2N$ (XXXVн, XXXVIи), Ar*) = 4-HO-3,5-(*tert*-$C_4H_9)_2C_6H_2$.

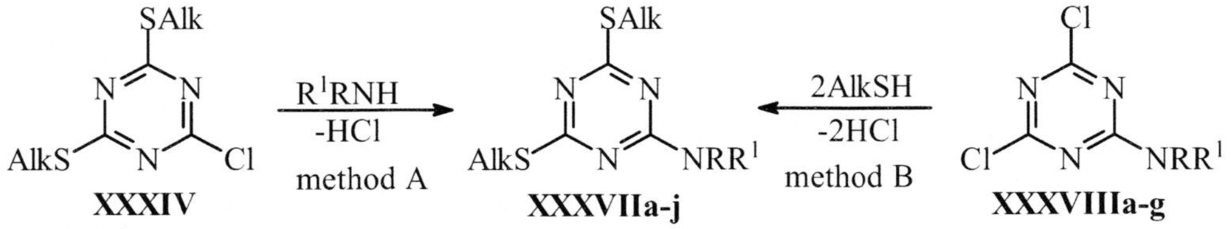

Alk = alkyl C_2-C_5;

R^1RN = $(CH_3)_2N$ (XXXVIIa, XXXVIIIa), *n*-$C_8H_{17}NH$ (XXXVIIb), *n*-$C_{18}H_{37}NCH_3$ (XXXVIIc, XXXVIIIb), Ar*)CH_2NH (XXXVIId), 4-ClC_6H_4NH (XXXVIIe), Ar*)NH (XXXVIIf, XXXVIIIc), morpholino (XXXVIIg, XXXVIIId), 2,2,6,6-$(CH_3)_4$-pyperidile-4-amino (XXXVIIh, XXXVIIIe), Et_2N (XXXVIIi, XXXVIIIf), thiazolyl-2-amino (XXXVIIj, XXXVIIIg).

Synthesis of N-substituted 2,4-diamino-6-alkylthio-*sym*-triazines by method B on the base of widespread industrial herbicides which are also N,N'-disubstituted 2,4-diamino-6-chloro-*sym*-triazines is of great interest from the point of broadening of biologically active substances assortment [48, 191]. Application of known herbicides of triazine type as initial substances such as 2,4-dithylamino-6-chloro- (XXXVIi; symazine), 2,4-di(isopropylamino)-6-chloro- (XXXVIj; propazine), 2-isopropylamino-4-ethylamino-6-chloro- (XXXVIk; atrazine) and 2-*tert*-butylamino-4-ethylamino-6-chloro-*sym*-triazine (XXXVIl; terbutylazine) and also 2-amino-4-anilino-6-chloro-*sym*-triazine (XXXVIm) in accordance with literature data [199] allows modifying of these preparations with obtaining of novel compounds possessing more valued properties probably. We established that 2-(R-amino)-4-(R^1-amino)-6-alkylthio-*sym*-triazines XXXIXa-e are formed with the yield 66-78% (Table 2.30) during boiling (for 12-16 hours) of equimolar amounts of chloro-*sym*-triazines XXXVIi-m and mercaptans mixture C_2–C_5 in DMFA in the presence of water solution of alakli:

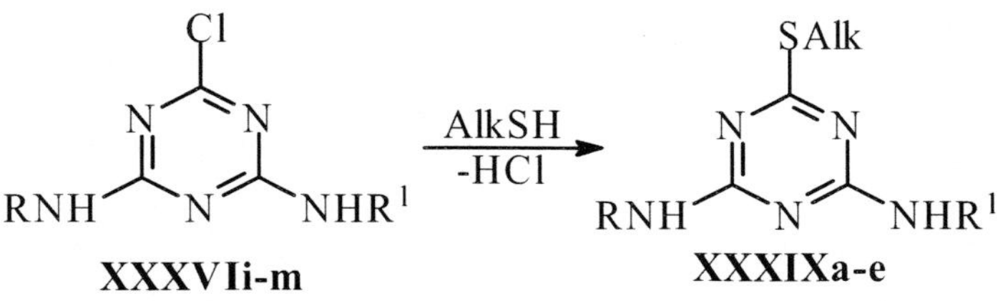

Alk = alkyl C_2-C_5; R = R^1 = C_2H_5 (XXXVIi, XXXIXa); R = R^1 = *iso*-C_3H_7 (XXXVIj, XXXIXb);R = *iso*-C_3H_7, R^1 = C_2H_5 (XXXVIk, XXXIXc); R = *tert*-C_4H_9, R^1 = C_2H_5 (XXXVIl, XXXIXd); R = H, R^1 = Ph (XXXVIm, XXXIXe).

Table 2.29. Constants, yield and data of elementary analysis of N-substituted 2-amino-4,6-dialkylthio-*sym*-triazines

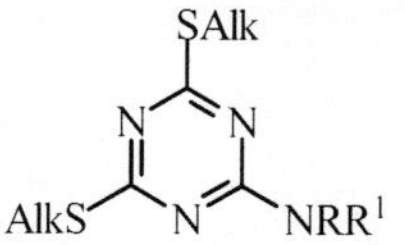

No.	Het	Yield, % (preparation method)	$T_{melt.}$, °C (solvent for crystallization)	n_D^{20}	S, mass %
XXXVIIa	$(CH_3)_2N$-	68(A), 74(B)	oil*)	1.5445	21.98
XXXVIIb	n-$C_8H_{17}NH$-	70(A), 66(B)	oil*)	1.5237	17.05
XXXVIIc	n-$C_{18}H_{37}(CH_3)N$-	66(A), 61(B)	42-49 (acetone)	-	16.76
XXXVIId	Ar*)CH_2NH-	72(A)	125-132 (dioxane-water,2:1)	-	13.48
XXXVIIe	4-ClC_6H_4NH-	58(A)	116-124 (ethanol-water, 2:1)	-	17.75
XXXVIIf	Ar*)NH-	72(A), 76(B)	89-98 (dioxane-water,2:1)	-	14.00
XXXVIIg	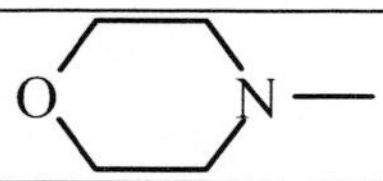	70(A), 63(B)	oil*)	1.5180	18.98
XXXVIIh	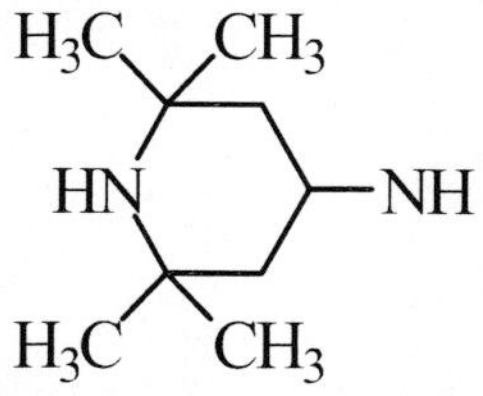	64(A), 69(B)	oil*)	1.5384	16.55
XXXVIIi	$(C_2H_5)_2N$-	66(A), 73(B)	oil*)	1.5438	19.92
XXXVIIj	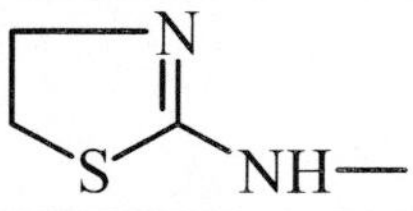	63(A)	104-115 (ethanol-water, 3:1)	-	34.09

Notes: *) Ar = 4-HO-3,5-(*tert*-Bu)$_2C_6H_2$;
**) Substances were purified by column chromatography on silicagel.

Table 2.30. Constants, yield and data of elementary analysis of N,N'-substituted 2,4-diamino-6-alkylthio-*sym*-triazines

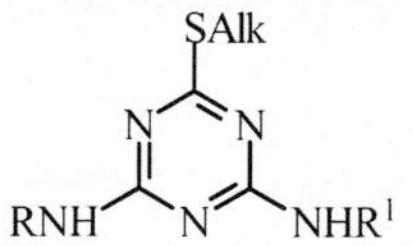

No.	R	R[1]	Yiled, %	$T_{melt.}$, °C (solvent for crystallization)	S, mass %
XXXIXa	C_2H_5	Et	75	72-78 (propanol-2)	12.07
XXXIXb	*iso*-C_3H_7	*iso*-C_3H_7	66	77-84 (DMFA-water, 3:1)	10.85
XXXIXc	C_2H_5	*iso*-C_3H_7	72	64-70 (propanol-2)	11.12
XXXIXd	C_2H_5	*tert*-C_4H_9	78	106-111 DMFA-water, 3:1)	10.74
XXXIXe	H	Ph	68	150-155 (dioxane-water, 5:1)	11.19

We also studied interaction of 2,4-dichloro-6-alkylthio- XXXIII and 2-chloro-4,6-dialkylthio-*sym*-triazines XXXIV with methanol, *p*-chlorophenol, 2,6-di-*tert*-butylhydroquinone and 4-hydroxydiphenylamine. 2,4-Dimethoxy-6-alkylthio- XXXXa and 2,4-diaryloxy-6-alkylthio-*sym*-triazines XXXXb-e with the yield 64-70% (Table 2.31) were obtained by reaction of 2,4-dichloro-*sym*-triazines XXXIII with sodium methylate, phenolate or thiolate (1 : 2 moles) under boiling (for 5 hours) in absolute methanol or aqueous dioxane (for 8-10 hours) (method A). 2-Methoxy-4,6-dialkylthio- XXXXIa and 2-aryloxy-4,6-dialkylthio-*sym*-triazines XXXXIb-d were synthesized with the yield 67-75% from 2-chloro-*sym*-triazines XXXIV.

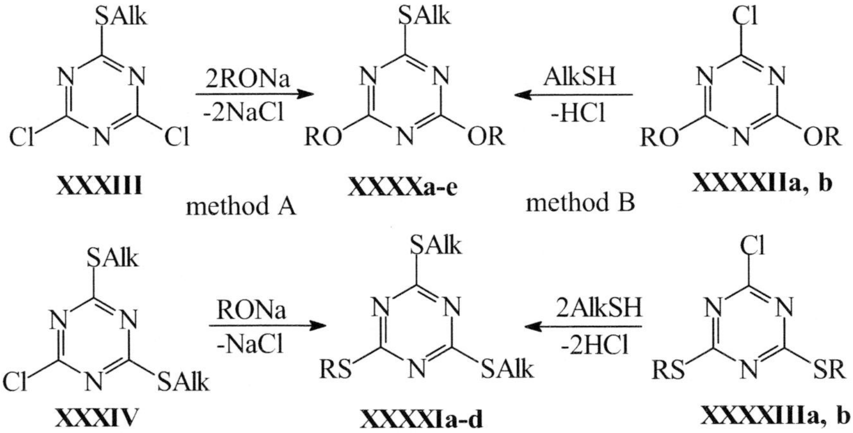

Alk = alkyl C_2-C_5; R = CH_3 (XXXXa, XXXXIa), 4-ClC_6H_4 (XXXXb, XXXXIb), benzthiazolyl-2- (XXXXe), 4-HO-3,5-(*tert*-$C_4H_9)_2C_6H_2$ (XXXXc, XXXXIc, XXXXIIa, XXXXIIIa), 4-$PhNHC_6H_4$ (XXXXd, XXXXId, XXXXIIb, XXXXIIIb).

Table 2.31. Constants, yield and data of elementary analysis of 2,4-di-(R-oxi)-6-alkylthio- and 2-(R-oxi)-4,6-dialkylthio-*sym*-triazines

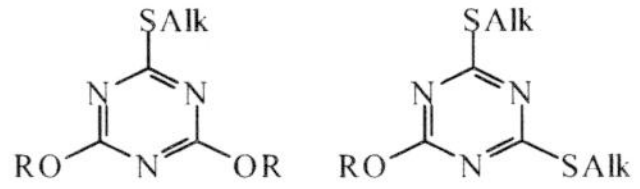

XXXXa-e XXXXIa-d

No.	R	Yield, % (preparation method)	$T_{melt.}$, °C (solvent for crystallization)	S, mass %
XXXXa	CH_3-	70(A)	40-47 (acetone-hexane, 3:1)	14.02
XXXXb	4-ClC_6H_4-	64(A)	88-94 (ethanol-water, 1:1)	6.95
XXXXc	Ar*)	68(A), 77(B)	140-146 (dioxane-water, 1:1)	5.08
XXXXd	4-$PhNHC_6H_4$-	65(A), 73(B)	120-125 (ethanol-water, 2:1)	5.84
XXXXe	Benzthiasolyl-2-	69(A), 72(B)	oil*) (n_D^{20} 1,5588)	19.34
XXXXIa	CH_3-	75(A)	oil*) (n_D^{20} 1,5368)	22.51
XXXXIb	4-ClC_6H_4-	67(A)	45-54 (benzene-hexane, 5:1)	17.42
XXXXIc	Ar*)	70(A), 68(B)	94-102 (toluene)	13.56
XXXXId	4-$PhNHC_6H_4$-	68(A), 72(B)	100-108 (ethanol-water, 1:1)	15.07
XXXXIVa	Ar*)	76(A), 74(B)	oil*) (n_D^{20} 1,5608)	14.97
XXXXIVb	Benzthiazolyl-2-	88(A), 85(B)	oil*) (n_D^{20} 1,5712)	34.92

Notes:*) Ar = 4-HO-3,5-(*tert*-Bu)$_2C_6H_2$; **) Substances were purified by column chromatography on silicagel.

The compounds XXXXc, d and XXXXIc, d were also obtained with good yield (Table 2.31) by alternative method – by interaction of 2,4-diaryloxy-6-chloro-*sym*-triazines XXXIIa, b or 2-aryloxy-4,6-dichloro-*sym*-triazines XXXXIIIa, b with mercaptans C_2–C_5 fraction (molar ratio 1 : 1 or 1 : 2) in aqueous dioxane in the presence of equivalent amount of alkali (method B).

Condensation of 2,4-dichloro-*sym*-triazines XXXIII during boiling (for 8-10 hours) with sodium salts of 4-hydroxy-3,5-di-*tert*-butylthiophenol or 2-mercaptobenzthiazole at their molar ratio 1 : 2 in aqueous dioxane leads to 2-alkylthio-4,6-di(R-thio)-*sym*-triazines XXXXIVa, b with the yiled 76-88% (method A). The same compounds were prepared by alternative technique – by boiling (for 6-8 hours) of 2,4- di(R-thio)-6-chloro-*sym*-triazines XXXXVa, b with equimolar amount of C_2–C_5 mercaptans mixture in aqueous dioxane (method B). We should note that the applied 2-chloro-4,6-disubstituted compounds XXXVIa-h, XXXXIIa, b, XXXXVa, b, and also 2,4-dichloro-6-substituted *sym*-triazines XXXVIIIa-f, XXXXIIIa, b in synthesizes mentioned above were preliminary obtained with the yield 74-92% on the base of cyanurchloride and corresponding amines, phenols or mercaptans in accordance with the methods reported in literature [200-204].

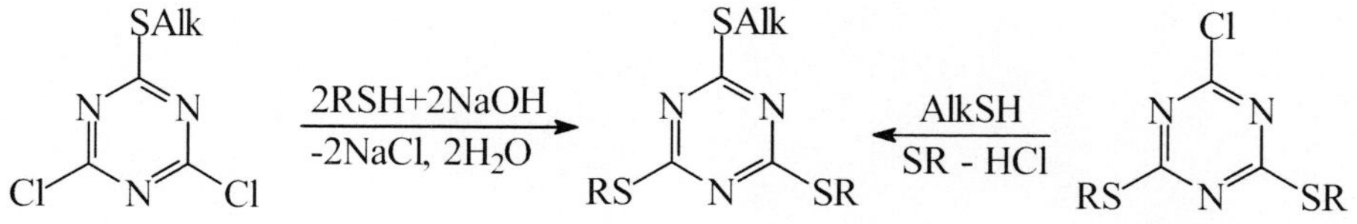

XXXIII **XXXXIVa-b** **XXXXVa, b**

Alk = alkyl C_2C_5; R = 4-HO-3,5-(*tert*-C_4H_9)$_2$$C_6H_2$ (**a**), benzthiazolyl-2 (**b**).

The data of Tables 2.28, 2.29 and 2.31 testify to close effectiveness of methods A and B, however the second one is more preferable for preparative needs since initial chloro- (XXXVIa-h, XXXXIIa, b, XXXXVa, b) and di-chloro-*sym*-triazines (XXXVIIIa-f, XXXXIIIa, b) may be easier purified (by recrystallization) in comparison with chloro-derivatives XXXIII and XXXIV. In other words it is advisable under synthesis of 2-alkyl-thio- and 2,4-dialkylthio-*sym*-triazines to introduce alkylthiole groups at final stage.

The constants and yield of synthesized sulfides of *sym*-triazine type are presented in Tables 2.28–2.31. The obtained compounds are solid substances with "expanded" melting temperatures or yellow viscous non-crystallizing oils well soluble in methylenechloride, ethers, acetone, alcohols, dioxane and aromatic hydrocarbons. Some of these compounds were purified by re-crystallization out of suitable solvents, and sulfides XXXVj, XXXVIIa, b, g, h, XXXXIa and XXXXVa were purified by column chromatography on silicagel.

The structures of synthesized derivatives of *sym*-triazine were confirmed by the data of IR and PMR-spectra. In IR spectra of these compounds in the region 1600-1000cm^{-1} there are absorption maximums of variable intensity characteristic for stretching and deformation vibrations of *sym*-triazine cycle. Intensive absorption maximums in the intervals 1565-1545, 1530-1520 and 1440-1425cm^{-1} should be assigned to stretching vibrations of *sym*-triazine cycle [183, 201-204]. We should note that the first mentioned bands in spectra of all obtained sulfides are slightly shifted to the short-wave region of spectrum in comparison with initial cyanurchloride (1510cm^{-1}).

The absorption maximums of variable intensity in the region 1115-1090 and 1020-1000cm^{-1} are conditioned by "breathing" vibrations of triazine cycle [183]. Intensive absorption bands at 815-800cm^{-1} practically independent on the nature of substituents in the cycle are characteristic for extra-planar deformation vibrations of cycle [183, 201, 202]. Frequencies of vibrations in the region 790-660cm^{-1} correspond to planar deformation vibrations of cycle [183, 203]. The number of bands of variable intensity in the regions 2700-2600, 1430-1420 and 800-600cm^{-1} presenting in spectra of all synthesized compounds should be assigned to stretching vibrations of C–S bond in accordance with literature data [167].

In spectra of 2,4-dichloro- XXXIII and 2-chloro-*sym*-triazines XXXIV there are intensive absorption maximums at 1300, 1260-1250, 1160 and 850-840cm^{-1} relating to stretching vibrations of C–Cl bond and characteristic for chloro-derivatives of *sym*-triazine [200]. There are no such frequencies in IR-spectra of other studied derivatives of *sym*-triazine.

Together with mentioned vibrations in IR spectra there also absorption maximums characteristic for vibrations of separate fragments in the cycle of *sym*-triazine. For example in high-frequency region of spectra of sulfides XXXVg, l, XXXVIIf, XXXXc, XXXXIc, and XXXXIVa containing fragments of spatially-hindered phenol there is narrow absorption band at 3655-3640cm^{-1} relating to stretching vibrations of HO- group in screened phenols [200, 205]. In spectra of these compounds in the interval 1265-1210cm^{-1} there are two absorption bands typical for bonds ArOH in spatially-hindered phenols [200] and also two groups of bands in the region 885-870 and 830-815cm^{-1} which should be assigned to non-planar deformation vibrations of C–H of tetrasubstituted benzene cycle in oxiaryl fragments [166, 167].

In spectra of N-substituted 2,4-diamino- (compounds XXXVb, d-i, k and XXXIXa-e) and 2-amino-*sym*-triazines XXXVIIb, d-f, h in the region of stretching vibrations of NH there is broad absorption band at 3340-3270cm^{-1}. Absorption maximums in the interval 1650-1580cm^{-1} relate to deformation vibrations of N–H bond in secondary amino-groups [166].

In IR-spectrum of 2-alkylthio-4,6-di(benzthiazolyl-2-thio)-*sym*-triazines XXXXIVb there are also absorption bands conditioned by benzthiazole fragment [183]: 1605, 1520, 1450, 1380 (stretching vibrations of condensed thiazole cycle), 1140, 1075 ("breathing" vibrations of the same cycle), 940 (γ-CH) and 790cm^{-1} (β-cycle).

Table 2.32. Individual composition of mixtures of synthesized sulfides of *sym*-triazines type (by the data of chromatography-mass spectrometry)

Alk	Content of components in compounds mixture, %		
	XXXIII	XXXIV	XXXXa
C_2H_5	3.0	2.9	2.8
n-C_3H_7	22.5	20.8	19.9
iso-C_3H_7	40.6	42.8	41.2
n-C_4H_9	5.5	5.7	5.4
iso-C_4H_9	22.5	22.0	25.2
n-C_5H_{11}	1.4	1.7	1.5
iso-C_5H_{11}	4.5	4.1	4.0

In PMR spectra of compounds XXXVg, l, XXXVIIf, XXXXc, XXXXIc and XXXXIVa signals of protons of hydroxyl groups are revealed in the shape of singlets in the region 5.05-5.14m.p. that is characteristic for screened phenols [205]. Protons signals of *tert*-butyl groups are observed in the shape of broadened singlets in the interval 1.52-1.64m.p.. Singlet signals at 7.10-7.22m.p. characterize two magneto-equivalent protons of 4-hydroxy-3,5-di-*tert*-butylphenyl fragments. In *sym*-triazines spectra XXXVb, d-i, k, XXXVIIb, d-f, h and XXXIXa-e the signals of protons of secondary amino-groups are presented in the shape of broadened singlets in the region 5.70-6.06m.p. characteristic or amino-derivatives of *sym*-triazine [201-203], diphenylamine [167] and pyperidine [183].

The distribution of individual components in their mixtures (Table 2.32) testifying to close reactivity of components in the initial mercaptans mixture was found for some derivatives of *sym*-triazine.

Chapter 3

SYNTHESIS AND TRANSFORMATIONS OF FUNCTIONALLY SUBSTITUTED ALKYLTHIOPROPANOLS AND ALKYLTHIOPROPIONIC ACIDS

3.1. SYNTHESIS AND TRANSFORMATIONS OF 1-ALKYLTHIO-3-CHLOROPROPANOLS-2

A lot of 1,3-disubstituted derivatives of propanol-2 reveal high and versatile biological activity and find an application as medicines [1], for example antiarhythmic and hypertensive preparation *Propranolol* – 1-isopropylamino-3-(1-naphthyloxi)propanol-2 or effective regulators of plants growth [7, 195] such as 1-alkoxi-3-pyperidinepropanols-2. Such compounds also serve as flotation reagents and reactive semi-products in thin organic synthesis [206].

So, various 3-substituted 1-alkylthio-propnaols-2, derivatives of 1-alkylthio-3-chloropropanols-2 synthesized in its turn by the interaction of mixture of natural mercaptans C_2–C_5 with 2-chloromethyloxirane (epichlorohydrine) represent great interest.

3.1.1. Interaction of Mercaptans with Epichlorohydrine

Some α-oxides (ethylene oxide, propylene oxide and epichlorohydrine) at present are available large-tonnage products as it was mentioned in Chapter 1 [19]. Their interaction with mercaptans is considered as one of the most interesting but insufficiently explored reaction. The data presented in literature [9, 36, 207, 208] are limited and contradictory, not allowing making the conclusions about conditions of reaction carrying out and nature of catalysts providing high selectivity and yield of products.

In accordance with modern conceptions [81] interaction of alkenes oxides (oxiranes) and compounds containing mobile H atom, including mercaptans is considered as reaction of nucleophilic addition by carbon atom of oxide's cycle:

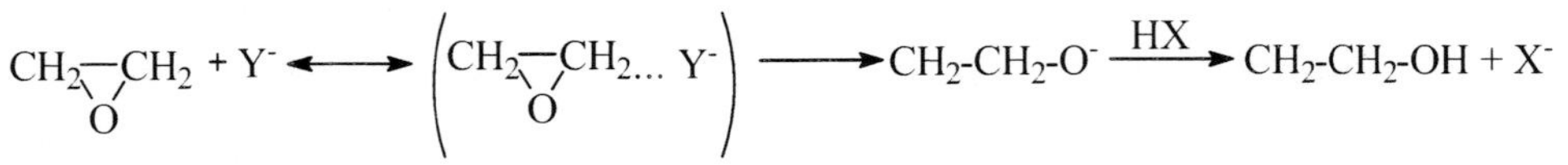

In general case reactions are carried out in the presence of both basic (hydroxides of alkali metals, aliphatic amines or pyridine) and acidic catalysts (mineral acids, Lewis acids, quaternary ammonium salts) [36, 207]. At that they note [81] that oxide cycle possessing high reactivity and lability is sensitive to any changes of reaction medium properties conditioning the specificity of mechanism and conditions of reactions in the case of one or another combination of their participants and also for the composition of reaction medium.

So, interaction of α-oxides with mercaptans in the presence of mineral or Lewis acids in contrast to catalysis in the presence of bases is accompanied by activation of oxide cycle:

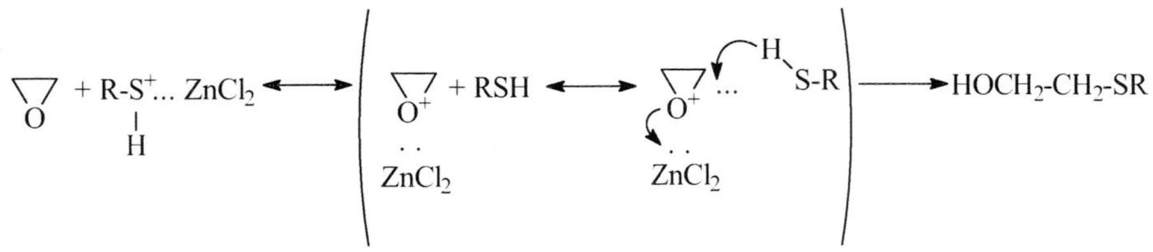

In our work we studied the influence of catalyst type on proceeding of the reaction of α-oxides with mercaptans by the example of interaction between epichlorohydrine and individual mercaptans: propanthiole-1, butanthiole-1 and heptanethiole-1 in the presence of basic (potassium hydroxide, amino-containing anionite AN-31) and acidic (boron trifluoride etherate, aluminus and zinc chlorides) catalysts.

Reaction of epichlorohydrine with propanthiole-1 was carried out in medium of inert solvent (hexane, petroleum-ether) at equimolar ratio of reagents and their initial concentration 9-12 vol. %. When we used anionite AN-31 (10 mass %) as catalyst at temperature 20-25°C for 10 hours there were no any changes in reaction mixture. Under the rise of reaction temperature up to 60°C and stirring for 3 hours conversion of initial reagents was only 3-5% with formation of mixture of 1-chloropropanol-2 XXXXVI, 3-chloro-2-propylthiopropanol-1 XXXXVII and 1-propylthio-3-chloropanthiole-2 XXXXVIIIa in the ratio 4 : 1 : 20 (by the data of gas-liquid chromatography). Further heating of reaction mixture, change of reagents ratio, increase of catalyst amount up to 20 mass %, and also addition of solvating solvent into reaction mixture (tetrahydrofuran, dioxane) did not increase the conversion of propanthiole-1.

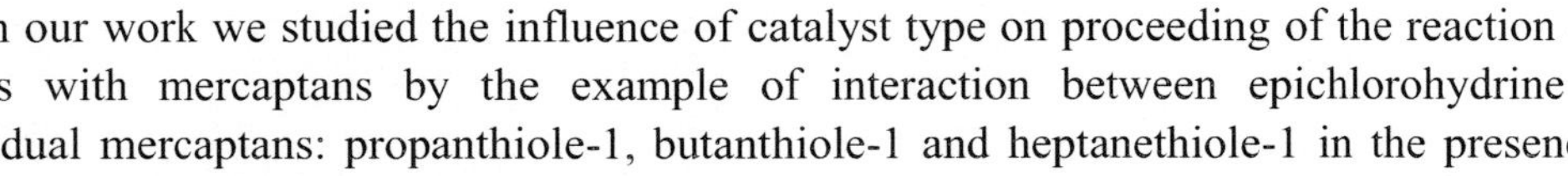

XXXXVI **XXXXVII** **XXXXVIIIa**

In the presence of 10 mass % of powder-like KOH at 25-30°C after 0.5 hour the conversion of initial substances reached 55-60% and yield of products (mixture of compounds XXXXVI, XXXXVII and XXXXVIIIa) was 45-50% at components ratio 1 : 4 : 26. The total conversion of initial propanthiole-1 and yield of products 76-80% is observed under the use of 35-40 mass % of KOH and molar ratio mercaptans : epichlorohydrine 1 : (1.2-1.25). In the case of thrice excess of propanthiole-1 at the same conditions the mixture of compounds

XXXXVII, XXXXVIIIa and also 1,3-di(propylthio)propanol-2 XXXXIX and 2,3-di(propylthio)propanol-1 L is formed with the ratio 2 : 17 : 1 : 25 and conversion of initial epichlorohydrine is 85-92%.

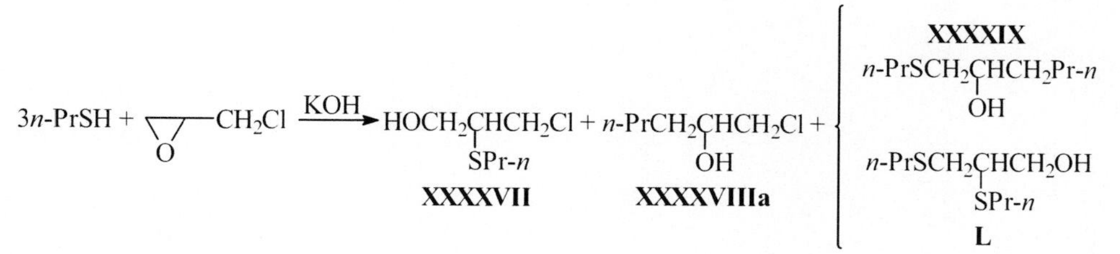

Interaction of epichlorohydrine with propanthiole-1 in the presence of Lewis acids mentioned above was studied at equimolar ratio of reagents. As a result of a number of experiments it was found that at temperature 25-30°C reaction was stopped at insignificant conversion of initial substances after 0.5 hours after addition of catalytic amounts of boron trifluoride etherate. In the reaction mixture together with initial substances there is the only reaction product 1-propythio-3-chloropropanol-2 XXXXVIIIa which yield is 405% and is not increased at further reaction proceeding, temperature rise, change of reagents ratio.

At analogous conditions under catalysis by $AlCl_3$ only insignificant conversion of reagents is also reached. Chloro-derivatives XXXXVI, XXXXVII and XXXXVIIIa and also di(propylthio)propanols XXXXIX and L in the ratio 1.5 : 1.2 : 1.0 : 2.5 : 1.7 with their total yield 5-6% were found in reaction mixture. Variation of reaction conditions does not increase reaction selectivity and yield of desired products.

It turned out that $ZnCl_2$ is the most effective catalyst in the presence of which the conditions providing complete conversion of propanthiole-1 and high selectivity in the formation of 1-propylthio-3-chloropropanol-2 XXXXVIIIa (90-95%) were determined. Reaction should be carried out in hexane or petroleum-ether solution at molar ratio propanthiole-1 : epichlorohydrine equal to 1 : (1.2-1.5), initial concentration of reagents 9-12%, temperature 20-25°C in the presence of 8-10 mass % of $ZnCl_2$ for 1 hour. The data on results of interaction between propanthiole-1 and epichlorohydrine at conditions mentioned above are presented in Table 3.1.

Then at analogous conditions 1-butylthio- XXXXVIIIb, 1-heptylthio- XXXXVIIIc and mixture of 1-alkylthio-3-chloropropanols XXXXVIIId were obtained with the yield 85-92% on the base of butanethiole-1, heptanethiole-1 and mercaptans C_2–C_5 fraction at conversion of the last ones equal to 93-97%.

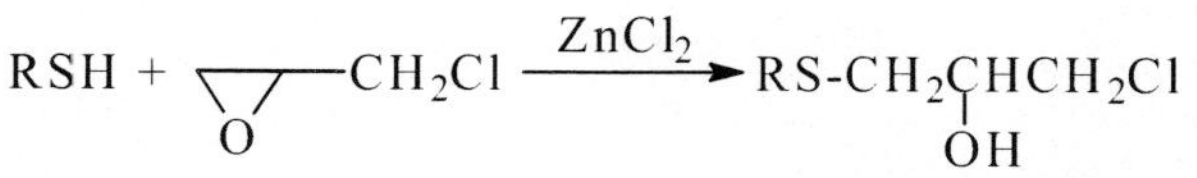

XXXXVIIIb-d

R = n-C_4H_9 (**b**), n-C_7H_{15} (**c**), alkyl C_2C_5 (**d**)

Physical-chemical characteristics of synthesized 1-alkylthio-3-chloropropanol-2 XXXXVIIIa-d are presented in Table 3.2. Constants of compounds XXXXVIIIa-c are in good agree with literature data, their structures were confirmed by spectral methods.

Table 3.1. The results of interaction between propanthiole-1 and epichlorohydrine at various conditions (solvent was hexane)

Molar ratio of reagents	Catalyst	Amount of catalyst, mass %	Temperature, °C	Reaction time, h	Reaction products and their ratio	Conversion of thioles,%	Total yield of products, %
1 : 1	-	-	60	20	Reaction does not proceed	-	-
1 : 1	AN-31	10	20	20	Reaction does not proceed	-	-
1 : 1	AN-31	10	60	40	XXXXVI : XXXXVII : XXXXVIIIa = 4 : 1 : 20	3-5	3-4
1 : 1	KOH solid	10	25	0.5	XXXXVI : XXXXVII : XXXXVIIIa = 1 : 4 : 26	55-60	45-50
1 : 1.25	KOH solid	35-40	25	3.0	XXXXVI : XXXXVII : XXXXVIIIa = 1 : 4 : 30	90	
3 : 1	KOH solid	10	25	0.5	XXXXVII : XXXXVIIIa: XXXXIX : L = 2 : 17 : 1 : 25	50-55	45-50
1 : 1	$(C_2H_5)_2O \cdot BF_3$	5	25	0.5	XXXXVIIIa	6-8	4
1 : 1	$AlCl_3$	5	25	0.5	XXXXVI : XXXXVII : XXXXVIIIa : XXXXIX:L = 1.5: 1.2 : 1 : 2.5 : 1.7	8-10	5-6
1 : 1.5	$ZnCl_2$	8-10	25	1.0	XXXXVIIIa	98	90-95

Table 3.2. Physical-chemical characteristics and yield of 1-alkylthio-3-chloropropanols-2 $RSCH_2CH(OH)CH_2Cl$

Compound	R	Yield, %	$T_{boil.}$, °C (millimeters of mercury)	n_D^{20}	d_4^{20}, g/cm^3
XXXXVIIIa	Ph	95	92-94 (2)	1.5005	1.0969
XXXXVIIIb	n-C_4H_9	92	96-98 (2.5)	1.4980	1.0938
XXXXVIIIc	n-C_7H_{15}	85	136-138 (2)	1.4934	1.0338
XXXXVIIId	C_2-C_5	87	80-110 (2)	1.4972	1.0945

In the IR spectra of these compounds there are absorption bands in the region 2975-2840cm^{-1} conditioned by stretching vibrations of C–H bonds in alkyl groups; they are revealed as a band with complex composition of structures: peaks at 2975-2960 and 2875-2865cm^{-1} relate to vibrations of methyl groups (antisymmetrical ν_{as} and symmetrical ν_s), and peaks at 2930-2920 and 2860-2840cm^{-1} relate to stretching vibrations of C–H in methylene and methyl groups (ν_{as} and ν_s) [167]. Absorption in the region 1400-1300 and 735-820cm^{-1} is conditioned by deformation vibrations of C–H. absorption band of variable intensity at 1460-1450cm^{-1} characterizes antisymmetrical deformation vibration of C–H in methyl groups (δ_{as}). Intensive band in the region of 1370cm^{-1} appearing as a result of symmetrical deformation of C–H bonds of methyl groups (δ_{as}) is stable in its position in spectra of all studied in given Section compounds.

In high frequency region of spectra of compounds XXXXVIIIa-d there are broad absorption bands at 3330-3200cm^{-1} characteristic for stretching vibrations of associated HO-groups in aliphatic alcohols [167]. The bands of average intensity in the region 1340-1315cm^{-1} characterize stretching vibrations of C–O bands [209].

PMR-spectra of compounds XXXXVIIIa-d also confirm their structures. Signals of protons of chloromethyl group are presented by doublets in the interval 3.54-3.67m.p. [167]. Signals of protons of CH groups bonded with hydroxyl are presented as asymmetrical multiplets in the region 3.28-3.90m.p. and signals of hydroxyl proton as broadened singlet at 5.52-5.80m.p..

When analyzing mass-spectra of 1-alkylthio-3-chloropropanols-2 XXXXVIIIa-d several general regularities of decomposition of molecules of these compounds under electron impact were revealed. Molecular masses of secondary alcohols XXXXVIIIa-c and of separate components of mixture of compounds XXXXVIIId were determined with the help of chromatography-mass spectrometry, and the character of further fragmentations was in good agree with presented structures. The primary fragmentation of molecular ions M^+ in mass-spectra of these compounds proceeds by several competitive directions the main of which are the following:

(a) elimination of $ClCH_2^\bullet$ fragment leading to formation of ion $[M\text{-}CH_2Cl]^+$;
(b) β-decomposition in relation to sulfur atom with formation of ions $R\text{-}S^+{=}CH_2$ and $ClCH_2CH{=}OH^+$;
(c) ejection of alkyl radical R [167, 210].
(d) Peaks corresponding to alkyl radicals in mass-spectra of studied compounds have the highest intensity.

3.1.2. Synthesis of 3-Substituted 1-Alkylthiopropanols-2

The mixture of 1-alkylthio-3-chloropropanols XXXXVIIId was used as key compounds for reception of various 3-substituted derivatives: N-substituted 3-amino-, 3-alkoxy- and 3-aryloxy-1-alkylthiopropanols-2, and also 3-alkylthio-2-hydroxypropyl esters of dithioaminoformic and dithiophosphoric acids.

N-substituted and N,N –disubstituted 1-alkylthio-3-aminopropanols-2 LIa-j were synthesized by interaction between mixture of chloroderivatives XXXXVIIId and available primary and secondary amines (diethylamine, butylamine, octylamine, N-methyloctadecylamine, benzylamine, aniline, 4-hydroxy-3,5-di-*tert*-butylaniline, 2-dimethylaminopropylamine, pyperidine and morpholine). We established that it is advisable to carry out the reactions at equimolar ratio of reagents in propanol-2 or dioxane at 50-60°C for 4-5 hours with the use of water solution of alkali as acceptor of HCl; at these conditions compounds LIa-j were obtained with the yield 76-88% (Table 3.3).

Table 3.3. Constants, yield and data of elementary analysis of N-substituted and N,N-disubstituted 1-alkylthio-3-aminopropanols-2 $AlkSCH_2CH(OH)CH_2NRR^1$

Compound	R^1RN	Yield, %	$T_{boil.}$, °C (millimeters of mercury) or $T_{melt.}$, °C (solvent for crystallization)	n_D^{20}	d_4^{20}, g/cm^3	S, mass %
LIa	$(C_2H_5)_2N$-	78	66-92 (2)	1.4854	0.9801	15.15
LIb	*n*-C_4H_9NH-	80	78-102 (4)	1.4892	0.9794	14.92
LIc	*n*-$C_8H_{17}NH$-	84	56-62 (ethanol-water, 1:1)	-	-	11.52
LId	*n*-$C_{18}H_{37}(CH_3)N$-	88	39-44 (MEK-heptane,4:1)	-	-	7.24
LIe	$PhCH_2NH$-	76	147-162 (2)	1.5034	1.0582	12.53
LIf	$(CH_3)_2N(CH_2)_3NH$-	77	104-126 (2)	1.4942	1.0345	12.84
LIg	PhNH-	75	80-88 (ethanol-water, 2:1)	-	-	14.08
LIh	4-HO-3,5-(*tert*-$C_4H_9)_2$-C_6H_2NH-	82	58-63 (benzene-hexane, 2:1)	-	-	9.05
LIi	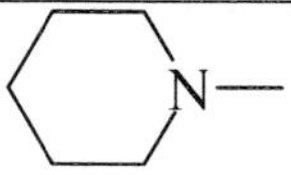	86	98-120 (2)	1.4965	0.9885	14.35
LIj	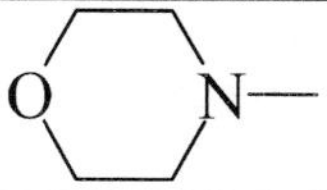	80	111-132 (2)	1.5002	1.0492	13.80

$$AlkSCH_2\underset{OH}{CH}CH_2Cl + R^1RNH \xrightarrow[-HCl]{} AlkSCH_2\underset{OH}{CH}CH_2NRR^1$$

XXXXVIIId **LIa-j**

Alk = alkyl C_2–C_5; R = R^1 = Et (a); R = Bu, R^1 = H (b); R = *n*-C_8H_{17}, R^1 = H (c); R = n-$C_{18}H_{37}$, R^1 = Me (d); R = $PhCH_2$, R^1 = H (e); R = $Me_2N(CH_2)_3$, R^1 = H (f); R = Ph, R^1 = H (g); R = 4-HO-3,5-(*tert*-Bu)$_2C_6H_2$, R^1 = H (h); R^1RN = piperidino (i), morpholino (j)

We synthesized 1-alkylthio-3-alkoxypropanols-2 LIIa-e with the high yield (Table 3.4) by the reaction of mixture of chloro-derivatives XXXXVIIId with sodium alcoholates in the excess of corresponding alcohol at 45-50°C for 8-10 hours. Such alkoxysubstituted LIIIa-c were obtained with the yield 84-88% by interaction of equimolar amounts of chlorides XXXXVIIId with sodium phenolate (generated on the base of corresponding phenols and sodium hydroxide in aqueous dioxane) for 3-4 hours at 50-60°C.

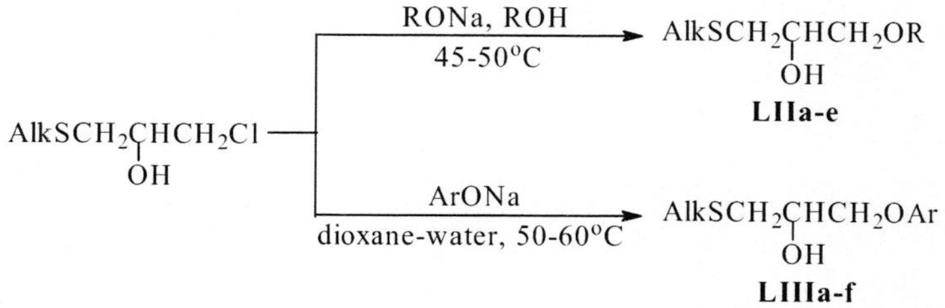

Alk = alkyl C_2-C_5; 52: R = Et (a), Bu (b), $PhCH_2$ (c), $Et_2NCH_2CH_2$ (d), tetrahydrofurfuryl (e);
53: Ar = Ph (a), 4-ClC_6H_4 (b), 4-HO-3,5-(*tert*-Bu)$_2C_6H_2$ (c).

Later on for some of synthesized β-aminoalcohols and 1-alkylthio-3-alkoxy(aryloxy)-propanols-2 the derivatives by the second HO-group were obtained. So, 3-substituted 1-alkylthio-2-(2-cyanethoxy)propanes LIVa-g were synthesized by cyanethylation of N,N-disubstituted β-aminoalcohols LIa, i, j, 3-alkoxy- LIIa, d and 3-aryloxypropanols-2 LIIIa, b:

$$\underset{\text{OH}}{\text{AlkSCH}_2\text{CHCH}_2\text{Z}} + \text{CH}_2\text{=CHCN} \xrightarrow[\Delta]{\text{CH}_3\text{ONa}} \underset{\text{OCH}_2\text{CH}_2\text{CN}}{\text{AlkSCH}_2\text{CHCH}_2\text{Z}}$$

LIa, i, j; LIIa, b **LIVa-g**

Alk = alkyl C_2-C_5; Z = Et_2N (LIa, LIVa), piperidino (LIi, LIVb), morpholino (LIj, LIVc), EtO (LIIa, LIVd), $Et_2NCH_2CH_2O$ (LIId, LIVe), PhO (LIIIa, LIVf) 4-ClC_6H_4O (LIIIb, LIVg).

Table 3.4. Constants, yield and data of elementary analysis of 1-alkylthio-3-alkoxy- and 1-alkylthio-3-alkyloxypropanols-2 $AlkSCH_2CH(OH)CH_2OR(Ar)$

Compound	R, Ar	Yield %	$T_{boil.}$, °C (millimeters of mercury) or $T_{melt.}$, °C (solvent for crystallization)	n_D^{20}	d_4^{20}, g/cm^3	S, mass %
LIIa	C_2H_5-	85	43-72 (1)	1.4654	0.9580	16.61
LIIb	*n*-C_4H_9-	88	88-112 (1)	1.4685	0.9621	14.54
LIIc	$PhCH_2$-	84	140-156 (1)	1.4732	1.0422	12.37
LIId	$(CH_3)_2NCH_2CH_2$-	88	111-130 (2)	1.4752	0.9708	11.88
LIIe	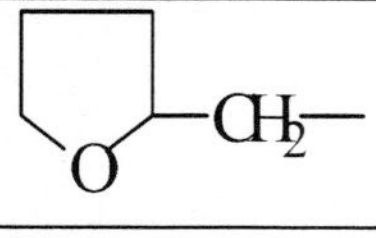	85	136-153 (2)	1.4694	0.9751	12.54
LIIIa	Ph	74	56-64 (ethanol-water,2:1)	-	-	13.12
LIIIb	4-ClC_6H_4	76	74-80 (propanol-2)	-	-	11.30
LIIIc	4-HO-3,5-(*tert*-$C_4H_9)_2$-C_6H_2	70	65-71 (ethanol-water,2:1)	-	-	8.41

It is obvious from Table 3.5 that the highest yields of β-cyanethyl esters LIVa-g are reached at molar ratio secondary alcohol : acrylonitrile equal to 1 : 3 and with the use of sodium methylate as catalyst. Cyanethylation of alcohols mentioned above was carried out at 45-50°C for 2-3 hours in the solution of waterless dioxane or tetrahydrofuran.

By boiling for 3-4 hours of N,N-disubstituted 1-alkylthio-3-aminopropanols-2 LIa, i, j and 3-alkoxypropanols-2 LIIa, c, e in acetic anhydride we obtained the corresponding 3-substituted 1-alkylthio-2-acetoxypropanes LVa-f with the yield 78-86%.

Table 3.5. Constants, yield and data of elementary analysis of 3-substituted 1-alkylthio-2-(2-cyanethoxy)propanes $AlkSCH_2CH(OCH_2CH_2CN)CH_2Z$

Compound	Z	Yield %	$T_{boil.}$, °C (millimeters of mercury) or $T_{melt.}$, °C (solvent for crystallization)	n_D^{20}	d_4^{20}, g/cm^3	S, mass %
LIVa	$(C_2H_5)_2N$-	73	128-138 (2)	1.4584	0.9602	10.64
LIVb	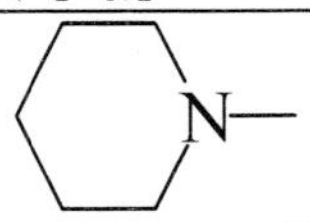	77	161-177 (2)	1.4677	1.0018	10.19
LIVc	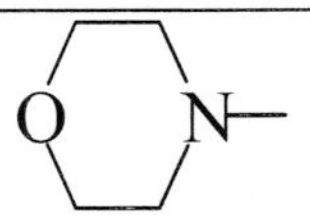	70	176-190 (1)	1.4691	1.0102	10.48
LIVd	C_2H_5O-	75	112-133 (1)	1.4708	1.0085	11.98
LIVe	$(C_2H_5)_2NCH_2CH_2O$-	64	186-198 (1)	1.4818	0.9874	9.51
LIVf	PhO-	68	69-75 (petroleum-ether)	-	-	9.88
LIVg	4-ClC_6H_4O-	72	64-72 (ethanol-water, 2:1)	-	-	9.54

Acylation of mentioned alcohols by benzoylchloride by Ph. Schott–E. Bauman's method in dioxane in the presence of alkali water solution leads to 3-substituted 1-alkylthio-2-benzoyloxypropanes LVIa-f with the good yield (Table 3.6):

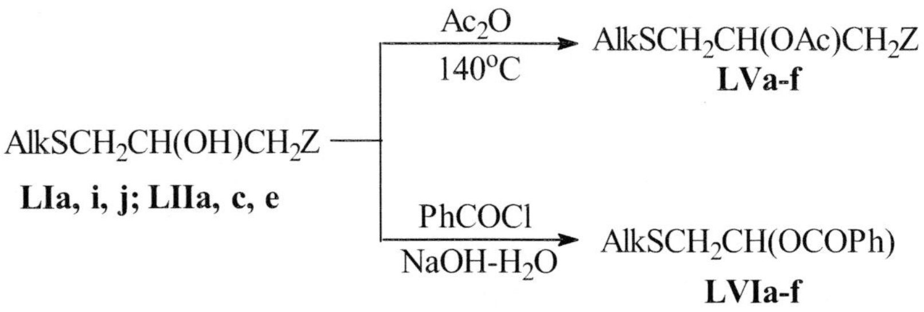

Alk = alkyl C_2-C_5; Z = Et_2N (LIa, LVa, LVIa), piperidino (LIi, LVb, LVIb), morpholino (LIj, LVc, LVIc), EtO (LIIa, LVd, LVId), $PhCH_2O$ (LIIc, LVe, LVIe), tetrahydrofurfuryloxy (LIIe, LVf, LVIf).

Ethers of phosphorous, dithioaminoformic and dithiophosphoric acids were synthesized on the base of 1-alkylthio-3-chloropropanols-2 XXXXVIIId and tested as additives for lubricating oils (Chapter 5).

Table 3.6. Constants, yield and data of elementary analysis of 3-substituted 1-alkylthio-2-acetoxy- and 1-alkylthio-2-benzoyloxypropanes $AlkSCH_2CH(OCOR)CH_2Z$

Compound	Z	R	Yield, %	$T_{boil.}$, °C (millimeters of mercury) or $T_{melt.}$, °C (solvent for crystallization)	n_D^{20}	d_4^{20}, g/cm³	S, mass %
LVa	$(C_2H_5)_2N$-	Me-	84	103-125 (1)	1.4502	1.0372	12.65
LVb	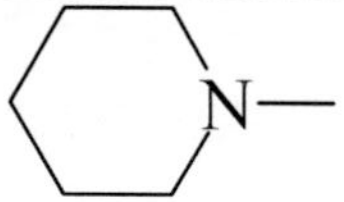	Me-	78	174-189 (1)	1.4588	1.0022	11.81
LVc	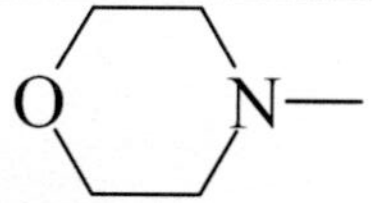	Me-	76	201-212 (1)	1.4609	0.9906	11.78
LVd	$(C_2H_5)O$-	Me-	86	130-147 (2)	1.4597	1.0010	14.08
LVe	$PhCH_2O$-	Me-	77	204-215 (2)	1.4618	1.0124	11.64
LVf	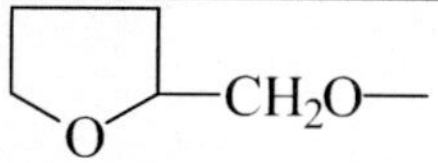	Me-	80	180-200 (1)	1.4704	1.0237	11.53
LVIa	$(C_2H_5)_2N$-	Ph-	72	58-66 (benzene-hexane,3:1)	-	-	9.27
LVIb	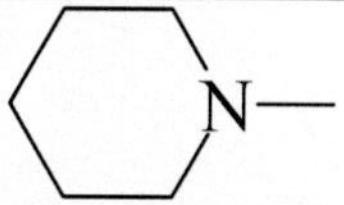	Ph-	75	84-91 (ethanol-water, 1:1)	-	-	8.65
LVIc	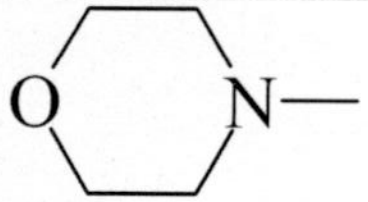	Ph-	67	94-102 (propanol-2)	-	-	8.52
LVId	C_2H_5O-	Ph-	78	65-72 (benzene)	-	-	10.04
LVIe	$PhCH_2O$-	Ph-	65	112-120 (propanol-2)	-	-	8.68
LVIf	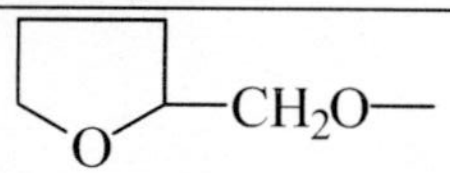	Ph	72	89-97 (ethanol-water, 3:1)	-	-	9.06

In accordance with literature data [168] phosphorus trichloride under its interaction with secondary alcohols forms the ethers of phosphorous acid. They established that under heating of 1-alkylthio-3-chloropropanols-2 XXXXVIIId with PCl_3 (3 : 1 moles) up to 70-75°C for 4-5 hours the corresponding ethers are also obtained:

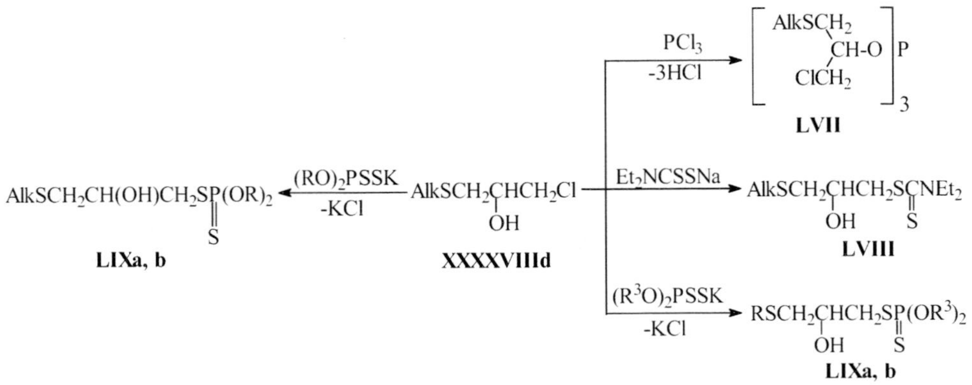

Alk = alkyl C_2-C_5; R = 4-*tert*-BuC_6H_4 (LXIa), 4-*tert*-$C_8H_{17}C_6H_4$ (LIXb)

As a result of condensation of chloro-derivatives XXXXVIIId with sodium diethyldithiocarbamate or potassium di(4-alkylphenyl)dithiophosphates in propanol-2 or dioxane (boiling for 12-15 hours) the ethers of diethyldithioaminoformic LVIII and dithiophosphoric LIXa, b acids are formed with high yield (75-90%). The obtained compounds LI–LIX represent viscous oils of light yellow or solid substances with "expanded" melting temperatures, well soluble in ether, acetone, alcohols, dioxane and aromatic hydrocarbons. Some of them were purified by distillation under vacuum or recrysatllization out of suitable solvents; ethers LVII, LVIII and LIXb were isolated from reaction mixtures and purified by column chromatography on silica-gel. Physical-chemical characteristics of compounds LI–LVI are presented in Tables 3.3–3.6, for compounds LVII, LVIII and LIXb – in Table 3.7.

Table 3.7. Physical-chemical properties of some N- and P-substituted 1-alkylthiopropanols-2

Compound	Substituents	$T_{boil.}$, °C (millimeters of mercury) or $T_{melt.}$, °C	n_D^{20}	d_4^{20}	S, mass %
LVII	P, Cl	-	1.5136	1.1721	17.08
LVIII	$(C_2H_5)_2NC(=S)S$-	-	1.5617	1.1054	33.02
LIXa	(4-*t*.C_4-$C_6H_4O)_2P(=S)S$-	145-170 (1)	1.5264	1.0921	18.18
LIXb	(4-*t*.C_8-$C_6H_4O)_2$-P(=S)S-	-	1.5452	1.0543	14.66

The structures of synthesized compounds were confirmed by the data of IR and PMR spectra. In IR spectra of products there are intensive absorption bands at 2935-2920 and 2860-2840cm^{-1} characterizing asymmetrical and symmetrical stretching vibrations of C–H in methylene groups [167] and a number of absorption bands of variable intensity in the interval 1465-1445cm^{-1} characteristic for asymmetrical and "cutter" deformation vibrations of CH_3– and CH_2–groups [167, 209].

In high-frequency region of spectra of compounds LIa-j, LIIa-e, LIIIa-c, LVIII and LIXa, b there are broad absorption bands at 3450-3320cm^{-1} characteristic for stretching vibrations of associated HO-groups in aliphatic alcohols [167, 211]. In IR-spectra of 0,05M solutions of β-aminoalcohols LIa-j in tetrachloroethylene (i.e. at conditions when there are no intermolecular associates [167]) positions of absorption bands at 3450-3400cm^{-1} are practically not changed and absorption maximums characteristic for non-associated HO-groups are absent. So, in the compounds LIa-j hydrogen of these groups participates mainly in the formation of intramolecular H-bond −O−H…N− [211]. In spectra of compounds LIb-h together with mentioned vibrations there are absorption maximums at 3315-3230 and 1650-1635cm^{-1} relating correspondingly to stretching and deformation vibrations of N−H in secondary amino-groups [167].

In spectra of cyanthiole ethers LIVa-g intensive absorption maximums in the region 2245-2235cm^{-1} characterize stretching vibrations of nitrile group [167, 209]. Intensive absorption bands at 1755-1740cm^{-1} correspond to carbonyl groups in esters LVa-f and LVIa-f [211].

In spectra of ethers of dithioaminoformic acid LVIII there absorption maximums of variable intensity at 1460, 1305 and 995cm^{-1} relating to vibrations of C=S bond in thioamide group. Absorption bands of average intensity in the regions 805-790 and 1050-1040cm^{-1} characteristic for vibrations of bonds P=S and P−Oar [167, 211] are revealed in spectra of compounds LIXa, b.

In PMR spectra of compounds LIa-j, LIIa-e, LIIIa-c, LVIII and LIXa, b the signals of hydroxyl protons are observed in the shape of broadened singlets at 5,42-5.80m.p.. The contents of individual components in some of synthesized derivatives of 1-alkylthiopropanol-2 were determined with the help of chromatography-mass spectrometry. The obtained values are differed because of decomposition of analyzing substances under electron impact with formation of additional amounts of propyl and ethyl (in the case of LVIII) fragments.

Table 3.8. The individual compositions of mixtures of synthesized derivatives of 1-alkylthiopropanol-2 (according to chromatography-mass spectrometry data)

Alk	The content of component in compounds mixture, %			
	XXXXVIIId	LIi	LIj	LIVb
Et	3.0	2.8	3.5	3.4
Pr	20.0	21.6	21.5	20.5
i-Pr	42.3	41.5	41.0	40.7
Bu	5.5	5.6	5.4	5.5
i-Bu	24.0	24.0	22.6	23.9
n-C_5H_{11}	1.2	1.4	1.5	1.5
i-C_5H_{11}	4.0	4.5	4.5	4.5

3.1.3. Synthesis of Derivatives of Oxazolidine on the Base of N-Monosubstituted 1-Alkylthio-3-Aminopropanols-2

As it is known [173, 212] condensation of β-aminoalcohols and their N-monosubstituted derivatives with carbonyl compounds is the most useful method of reception of 2-mono- and

2,3-disubstituted oxazolidines. We studied the interaction of N-monosubstituted 1-alkylthio-3-aminopropanols-2 LIb, c, e-h with aliphatic and aromatic aldehydes.

We showed that it was advisable to carry out the cyclocondensation of β-aminoalcohols LI b, c, e-h with paraform, isobutyric aldehydes, benzaldehyde, 4-hydroxy-3,5-di-*tert*-butylbenzaldehyde and furfural under boiling (4-7 hours), at equimolar ratio of reagents in waterless benzene and azeotrope removal of formed water. At these conditions the desired 3-substituted LXa-e and 2,3-disubstituted LXf-u are formed with the yield 60-84% (Table 3.9).

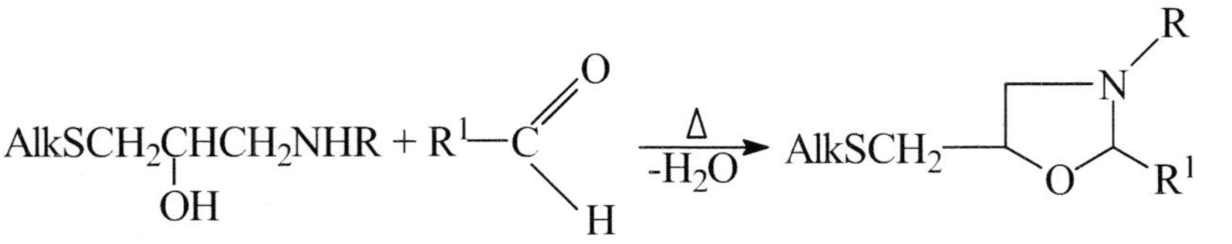

LIb, c, e-h **LXa-u**

Alk = alkyl C_2-C_5; R = Bu (LIb, LXa, f, j, n, s), C_8H_{17} (LIc, LXb, k, o), $PhCH_2$ (LIe, LXc, l, p), $Me_2N(CH_2)_3$ (LIf, LXg, q), Ph (LIg, LXg, h, t), 4-HO-3,5-(*tert*-Bu)$_2C_6H_2$ (LIh, LXe, i, m, r, u); R^1 = H (LXa-e), *i*-Pr (LXf-i), Ph (LXj-m), 4-HO-3,5-(*tert*-Bu)$_2C_6H_2$ (LXn-r), furyl-2 (LXs-u).

The structures of synthesized compounds are confirmed by the data of IR- and PMR spectroscopies and elementary analysis. In the IR spectra of heterocycles LXa-u the absorption maximums of variable intensity characterizing vibrations of oxazolidine fragments are revealed [183, 213]: 1480-1465, 1445-1420 ("cutter" vibrations of CH_2), 1190-1175, 1160-1145, 1120-1105 (stretching vibrations of aromatic cycle), 950-935 ("breathing" vibrations of the same cycle) and 850-830cm^{-1}. Absorption bands at 2700-2600, 1430-1410 and 800-670 cm^{-1} characteristic for vibrations of bond [167] are also observed in spectra. In spectra of LXe, i, m-r, u there is also narrow intensive band at 3640-3620cm^{-1} relating to stretching vibrations of HO-group in spatially-hindered phenols [205].

In PMR spectra of synthesized compounds there are several groups of signals differing in values of chemical shift. Signals of protons of CH_2-groups are presented by doublets and triplets in the region 1.02-1.16 m.p.. Complex multiplets in the interval 1.22-1.46m.p. should be assigned to protons of methylene units, in the interval 1.72-1.90m.p. – to protons of CH-groups, and signals shifted to weak field (3.28-3.76m.p.) – to protons of CH_2- and CH-groups bonded with sulfur atom.

The signal of 5-H proton in spectra of oxazolidines LXa-u is observed as non-symmetrical multiplet in the interval 4.25-4.74m.p.. Protons of methylene groups in position 2in spectra of 3,5-disubstituted oxazolidines LXa-e relate to typical AB-system of two non-symmetrical doublets at 3.92-4.18 and 4.25-4.38m.p. with the constants of spin-spin coupling J_{AB} = 15.0-17.6Hz [167].

In spectra of compounds LXe, i, m-r, u containing 4-hydroxy-3,5-di-*tert*-butylphenyl groups the signals of protons of hydroxyl groups are revealed as singlets in the region 5,05-5.15m.p. that is characteristic for screened phenols [205].

The contents of individual components in some of synthesized oxazolidines were determined with the help of chromatography-mass spectrometry (Table 3.10).

Table 3.9. Constants, yield and data of elementary analysis of 2-R^1-3-R-5-alkylthiomethyloxazolidines

Compound	R	R^1	Yield, %	$T_{boil.}$, °C (millimeters of mercury) or $T_{melt.}$, °C (solvent for crystallization)	n_D^{20}	d_4^{20}, g/cm^3	S, mass %
LXa	*n*-C_4H_9-	H-	69	72-95 (4)	1.4784	0.9433	14.82
LXb	*n*-C_8H_{17}-	H-	64	111-132 (3)	1.4835	0.9740	11.61
LXc	$PhCH_2$-	H-	65	119-140 (2)	1.4915	1.0234	13.04
LXd	Ph-	H-	77	37-44 (benzene-hexane, 3:1)	-	-	13.75
LXe	Ar*$^)$-	H-	72	42-51 (petroleum-ether)	-	-	12.62
LXf	*n*-C_4H_9-	*i*-Pr-	71	68-92 (1)	1.4874	1.0115	12.71
LXg	$(CH_3)_2N(CH_2)_3$-	*i*-Pr-	60	78-102 (3)	1.5035	0.9634	11.63
LXh	Ph-	*i*-Pr-	80	62-68 (petroleum-ether)	-	-	11.67
LXi	Ar*$^)$-	*i*-Pr-	76	67-75 (petroleum-ether)	-	-	8.18
LXj	*n*-C_4H_9-	Ph-	62	128-147 (1)	1.5084	1.0841	11.04
LXk	*n*-C_8H_{17}-	Ph-	65	150-168 (2)	1.5108	1.0951	9.63
LXl	$PhCH_2$-	Ph-	78	42-50 (petroleum-ether)	-	-	10.21
LXm	Ar*$^)$-	Ph-	83	74-82 (ethanol-water, 1:1)	-	-	7.71
LXn	*n*-C_4H_9-	Ar*$^)$-	74	38-46 (petroleum-ether)	-	-	7.62
LXo	*n*-C_8H_{17}-	Ar*$^)$-	80	56-65 (isooctane)	-	-	7.02
LXp	$PhCH_2$-	Ar*$^)$-	73	87-94 (ethanol-water, 2:1)	-	-	7.01
LXq	$(CH_3)_2N(CH_2)_3$-	Ar*$^)$-	71	40-47 (isooctane)	-	-	7.34
LXr	Ar*$^)$-	Ar*$^)$-	77	93-101 (benzene-hexane, 3:1)	-	-	5.93
LXs	*n*-C_4H_9-	furyl-2-	63	95-103 (2)	1.5085	1.0912	11.51
LXt	Ph-	furyl-2-	84	64-72 (ethanol-water, 3:1)	-	-	9.94
LXu	Ar*$^)$-	furyl-2-	81	60-68 (benzene-heptane, 2:1)	-	-	7.75

Note: *$^)$ Ar = 4-HO-3,5-(*tert*-C_4H_9)$_2$$C_6H_2$-

Table 3.10. The individual compositions of mixtures of synthesized oxazolidines (according to chromatography-mass spectrometry data)

Alk	The content of component in compounds mixture, %		
	LXc	LXi	LXt
Et	2.9	3.0	3.3
Pr	22.1	20.5	21.9
i-Pr	41.8	41.5	41.7
Bu	5.5	5.7	5.6
i-Bu	21.6	23.0	21.4
n-C_5H_{11}	1.5	1.3	1.4
i-C_5H_{11}	4.6	5.0	4.7

3.2. Synthesis and Transformations of Functional Derivatives of B-Alkylthiopropionic Acids

Earlier in Section 2.2.2 we described the reception and chemical transformations of alkylthioacetic acids and their functional derivatives. The logical continuation of this direction is the development of preparative methods of synthesis of functional derivatives of β-alkylthiopropionic acids which except independent importance represent essential interest as semi-products for preparation of nitrogen-containing heterocycles with β-alkylthioethyl substituents.

3.2.1. Synthesis of Functional Derivatives of B-Alkylthiopropionic Acids

Initially we developed preparative methods of reception of the most available derivatives of mentioned acids – nitriles LXI and esters LXIIa, b with the use of mercaptans C_2–C_5 mixtures as key compounds. As it is known [214, 215] it is convenient to synthesize the compounds including β-cyanethyl or β-alkoxycarbonylethyl groups by A. Mikhael's reaction, namely by interaction of compounds containing mobile hydrogen atom with acrylonitrile or ethers of acrylic acid. As a result of a number of experiments we established that cyanethylation of mercaptans C_2–C_5 mixture led to corresponding β-alkylthiopropionitriles LXI with the yield 86-92%. The conditions of reaction are: molar ratio mercaptans : acrylonitrile = 1 : 1,25, temperature 20-40°C, reaction time 1,5-2 hours, presence of catalytic amount of sodium methylate.

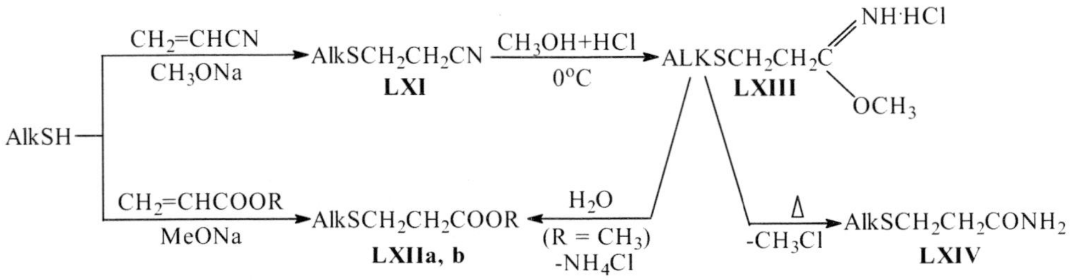

Alk = alkyl C_2-C_5; R = CH_3 (a), C_2H_5 (b)

Under the use of other catalysts of basic type (water solution of alkali, piperidine, benzyltrimethylammonium hydroxide, triethylamine) in this reaction the yield of desired products is decreased down to 73-76%.

Methyl LXIIa and ethyl LXIIb esters of β-alkylthiopropionic acids were synthesized by interaction of equimolar amounts of the same mercaptans mixture with ethers of acrylic acid at 30-35°C for 3-4 hours in the presence of sodium methylate with the yield 80-84%.

The obtained nitriles LXI and esters LXIIa, b are high boiling colorless or light yellow liquids, well soluble in the majority of organic solvents. the structured of these compounds were confirmed by the data of OR- and PMR-spectra and also by data of elementary nalysis.

In IR spectrum of adducts LXI there is intensive absorption maximum at 2240cm^{-1} conditioned by vibrations of cyan-group [167]. In spectra of esters LXIIa, b the intensive absorption bands in the region 1715-1710cm^{-1} characterize stretching vibrations of carbonyl group [167, 209]. In PMR spectra of mentioned compounds the signals of protons of ethylene fragments are revealed as non-symmetric triplets at 3.18-3.33m.p. (-CH_2S-) and at 2.72m.p. (-CH_2CN) or 2.88-3.02m.p. (-CH_2COOR) [167].

One of the most interesting transformations of nitriles of carbonic acids is the interaction with alcohols and hydrogenhalogens – G. Pinner's reaction [216]. The forming hydrochlorides of iminoethers (imidates) possess high reactivity and due to this fact the compounds of various classes may be obtained – esters, amides, *orto*-esters, amidines, amidrazones, etc., and also five- and six-membered nitrogen-containing heterocycles with several heteroatoms (imidazolines, benzazoles, 1,3,4-oxadiazoles, 1,2,4-triazoles, pyrimidines, *sym*-triazines, etc.) [174].

Only two works reported about the synthesis on the base of individual thioles of hydrochloride of iminoethers of β-ethylthio- [217], β-*iso*-butylthio- and β-benzylthiopropionic [218] acids before we began our work. That is why we carried out systematic investigation of G. Pinner's reaction on the base of nitriles mixture LXI. We established that when passing the current of dry HCl and absolute methanol in waterless ether through equimolar mixture of nitriles LXI at 0-5°C for 1-2 hours with further ageing of reaction mixture at 20°C for 24 hours the hydrochlorides of methyl iminoethers of β-alkylthiopropionic acids LXIII were formed with high yield (85-90%). When the molar ratio nitriles LXI : methanol is changed up to 1 : 2–1 : 2.5 the yield of compounds LXIII is decreased down to 70-73%.

The obtained salts of iminoethers LXIII are stable compounds and may be kept for a long time without access of moisture. In their IR-spectrum there are characteristic absorption bands at 1675-1670cm^{-1} relating to stretching vibrations of bond $C{=}N^+$ [166]. Absorption bands of average intensity in the interval 1025-1020cm^{-1} should be assigned to stretching vibration of C–O bond in iminoether group. In the high-frequency region of spectrum stretching vibrations of group $C{=}N^+{-}H$ are presented by absorption maximums at 3130-3120cm^{-1} characteristic for imine salts [219].

Structures of synthesized hydrochlorides of iminoethers LXIII were confirmed also by their transformations into other functional derivatives of β-alkylthiopropionic acids – amides LXIV and methylethers LXIIa. Thermal decomposition of hydrohalogenides of iminoethers of carbonic acids with formation of amides and alkylhalogenides as it is known [174] serves as their characteristic reaction usually called as G. Pinner's regrouping. We found that compounds LXIII under their heating until boiling are decomposed during several minutes forming amides of β-alkylthiopropionic acids LXIV with the yield 88-91%.

Hydrolysis of iminoethers hydrochlorides being suitable preparative method of reception of esters of various structures at the same time is one the simplest and reliable methods of their identification [174]. Actually hydrolysis of compounds LXIII at 20°C for 0.5 hour leads to formation of methylethers of β-alkylthiopropionic acids LXIa with the yield 80-82% and identical to those ones obtained by interaction of mercaptans C_2–C_5 mixture with methylacrylate.

Hydrazides of carbonic acids (N-acylhydrazines) traditionally attract attention of researches by their high reactivity and broad synthetic possibilities [168, 220, 221]. As key compounds we used hydrazides LXV of corresponding acids which were obtained by boiling (6-8 hours) of methyl ethers LXIIa with 80% hydrazinehydrate (molar ratio 1 : 1.5) in ethanol with the yield 85-88% for synthesis of some nitrogen-containing heterocycles with β-alkylthioethyl substituents.

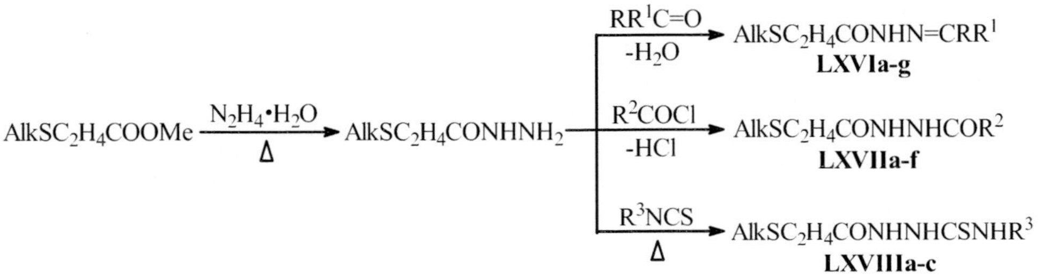

Alk = alkyl C_2-C_5;

66: R = *iso*-C_3H_7 (a), Ph (b), 4-HO-3,5-(*tert*-C_4H_9)$_2C_6H_2$ (c), 5-nitrofuryl-2 (d), R^1 = H; R = R^1 = CH_3 (e); R = Ph, R^1 = CH_3 (f); R + R^1 = -$(CH_2)_4$- (g);

67: R^2 = Ph (a), 4-$NO_2C_6H_4$ (b), $PhCH_2$ (c), $AlkSCH_2$ (d), 4-HO-3,5-(*tert*-C_4H_9)$_2C_6H_2SCH_2$ (e), thienyl-2 (f);

68: R^3 = CH_2=$CHCH_2$ (a), Ph (b), 4-ClC_6H_4 (c)

N-(β-alkylthiopropionyl)hydrazones of aldehydes LXVIa-d and ketones LXVIe-g were obtained with high yield by boiling (5-6 hours) of equimolar amounts of hydrazides LXV and corresponding aldehydes (2-methylpropanal, benzaldehyde, 4-hydroxy-3,5-di-*tert*-butylbenzaldehyde, 5-nitrofurfurol) or ketone (acetone, acetophenone, cyclopentanone) in solution of propanol-2 in the presence of catalytic amount of acetic acid (Table 3.11).

Acylation of hydrazines LXV by chloroanhydrides of various carbonic acids under boiling (for 2-4 hours) in pyridine solution leads to N-(β-alkylthiorpopionyl)-N'-acylhydrazines LXVIIa-f with the yield 68-85% (Table 3.12). By reaction of hydrazides LXV with allyl-, phenyl- and 4-chlorophenyl-*iso*-thiocyanate the 1-(β-alkylthiopropionyl)-4-R^3-thiosemicarbazides LXVIIIa-c were synthesized and their best yields were reached at equimolar ratio of reagents at conditions of boiling in propanol-2 solution.

The structures of synthesized hydrazides LXV and their derivatives LXVI-LXVIII were confirmed by data of IR and PMR spectroscopies and elementary analysis. In IR spectra of all compounds stretching vibrations of C–S bonds are characterized by absorption maximums of variable intensity in the region 1215-1165cm^{-1} [167]. In spectra of hydrazides LXV in the region 3400-3180cm^{-1} there are absorption maximums of NH-groups entering into intermolecular associates.

Table 3.11. Constants, yield and data of elementary analysis of N-(β-alkylthiopropionyl)hydrazones of carbonyl compounds $AlkSCH_2CH_2CONHN{=}CRR^1$

Compound	R	R^1	Yield, %	$T_{melt.}$, °C (solvent for crystallization)	S, mass %
LXVIa	*iso*-C_3H_7-	H-	77	82-88 (ethanol-H_2O, 2:1)	13.84
LXVIb	Ph-	H-	83	137-143 (benzene)	12.88
LXVIc	4-HO-3,5-(*tert*-$C_4H_9)_2C_6H_2$-	H-	80	61-66*)	8.66
LXVId	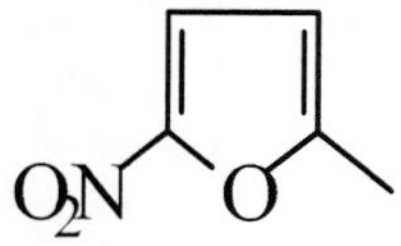	H-	75	124-130 (dioxane-water, 1:1)	10.52
LXVIe	CH_3-	CH_3-	84	oil*) (n_D^{20} 1,5489)	15.74
LXVIf	Ph-	CH_3-	86	81-85 (petroleum-ether)	11.28
LXVIg	-$(CH_2)_4$-		82	142-148 (ethanol)	13,07

Note: *) The compounds were purified by column chromatography on silica gel

Table 3.12. Constants, yield and data of elementary analysis of N-(β-alkylthiopropionyl)-N'-acylhydrazines (LXVIIa-f) $AlkSCH_2CH_2CONHNHCOR^2$ and 1-(β-alkylthiopropionyl)-4-R^3_thiosemicarbazides (LXVIIIa-c) $AlkSCH_2CH_2CONHNHCSNHR^3$

Compound	R^2	R^3	Yield,%	$T_{melt.}$, °C (solvent for crystallization)	S, mass %
LXVIIa	Ph-	-	74	133-139 (ethanol-water, 1:1)	11.44
LXVIIb	4-$NO_2C_6H_4$-	-	85	150-155 (ethanol)	10.70
LXVIIc	$PhCH_2$-	-	77	72-77 (ethanol-water, 1:1)	10.62
LXVIId	4-HO-3,5-(*tert*-$C_4H_9)_2C_6H_2SCH_2$-	-	72	102-109 (petroleum-ether)	14.68
LXVIIe	$AlkSCH_2$-	-	68	58-67 (petroleum-ether)	22.11
LXVIIf	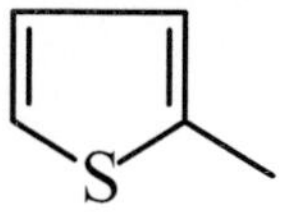	-	78	138-143 (ethanol-water, 1:1)	22.45
LXVIIIa	-	$CH_2{=}CHCH_2$-	82	142-147 (propanol-2)	24.21
LXVIIIb	-	Ph-	77	160-166 (dioxane-water, 2:1)	21.48
LXVIIIc	-	4-ClC_6H_4-	75	172-178 (propanol-2)	18.47

Intensive absorption band at 1650cm^{-1} may be assigned analogously to secondary amides to stretching vibrations of carbonyl group (band "amide I") [167]. Such decrease of frequency of stretching vibrations of group C=O in comparison with initial ethers ($\nu_{c=o}$ 1715cm^{-1}) is characteristic for N-acylhydrazines [220]. In spectrum there is also a number of absorption bands in the region 1580-1565cm^{-1} conditioned by deformation vibrations of NH and in accordance with [221] bands at 1140-1135cm^{-1} relate to stretching vibrations of N–N-bond.

In spectra of compounds LXVIa-g there are intensive absorption bands in the region 1650-1635cm^{-1} characteristic for stretching vibrations of group C=N in N-acylhydrazones [222, 223] and also in the region 1690-1670cm^{-1} characteristic to vibrations of the type "amide I" of N-acyl group [167]. In spectra of thiosemicarbazide derivatives LXVIIIa-c there is a number of absorption bands of variable intensity at 1550-1540, 1460-1450, 1435-1420 and 1330-1320cm^{-1} which should be assigned to stretching vibrations of C=S group in thioamide fragment [167, 224].

In PMR spectra of compounds LXV-LXVIII the signals of protons of secondary NH-groups are revealed as broadened singlets in weak field at 9.90-10.42m.p.. In spectra of N-acylhydrazones of aldehydes LXVIa-d signals of protons of CH=N groups are observed as broadened singlets at 7.45-7.78m.p.

The individual compositions of mixtures of some synthesized derivatives of β-alkylthiopropionic acids established by chromatography-mass spectrometry are presented in Table 3.13. The observing fluctuations in compositions of studied substances are in good agree with regularities of fragmentation of components of compounds mixture under electron impact.

Except compounds reported above in this Section we present data on synthesis of S-alkyl ethers of thiocarbonic acids. It is known [10, 79] that S-alkyl ethers of thiocarbonic acids (thiole ethers of carbonic acids) may be produced by acylation of mercaptans by chloroanhydrides of carbonic acids in the presence of bases (water solutions of alkali, pyridine, dimethylaniline, etc.). However these reactions are often accompanied by resinification and formation of by-products that significantly decreases the yield of desired products.

In this work we used another method for reception of thiol ethers consisting in hydrolysis of reactive salts of iminoethers [225]. Thioiminoethers hydrochlorides of carbonic acids were synthesized by G. Pinner's reaction [216, 226] – under passing of a current of dry HCl (gas) for 1-1.5 hours through equimolar mixture of corresponding nitrile and mercaptans C_2–C_5 mixture in waterless ether or chloroform 0-5°C.

Table 3.13. The individual compositions of mixtures of synthesized derivatives of β-alkylthiopropionic acids (according to chromatography-mass spectrometry data)

Alk	The content of component in compounds mixture, %					
	LXI	LXIIa	LXV	LXVIa	LXVIIa	LXVIIIb
C_2H_5	3.4	2.9	3.5	3.5	3.2	4.0
n-C_3H_7	20.7	21.5	21.0	19.8	21.6	21.5
iso-C_3H_7	41.3	40.0	39.8	40.5	40.2	40.5
n-C_4H_9	5.5	5.8	5.6	5.6	5.4	5.5
iso-C_4H_9	22.6	23.7	24.5	25.1	23.6	22.5
n-C_5H_{11}	1.8	1.7	1.5	1.6	1.5	1.7
iso-C_5H_{11}	4.7	4.4	4.1	3.9	4.5	4.3

Intermediate thioiminoethers hydrochlorides are treated by water at 20°C without isolation out of reaction mixtures; at these conditions the desired S-alkyl ethers of thiocarbonic acids LXIXa-j are formed with the yield 74-88% (Table 3.14).

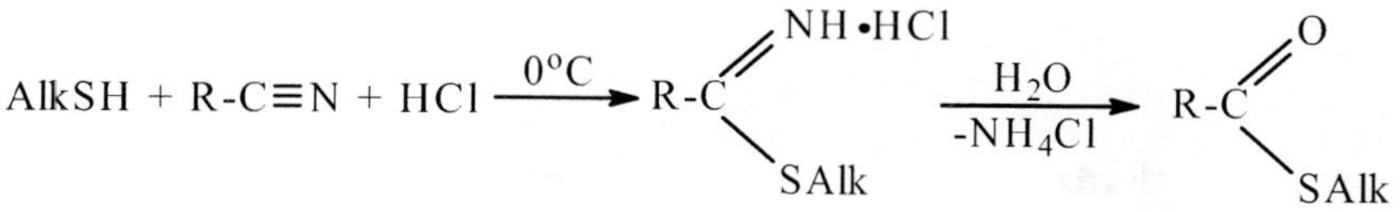

LXIXa-j

Alk = alkyl C_2-C_5; R = n-C_5H_{11} (a), n-$C_{11}H_{23}$ (b), n-$C_{17}H_{35}$ (c), Ph (d), naphtyl-1 (e), inolyl-3-methyl (f), pyridile-3 (g), benzthiazolyl-2-thiomethyl (h), β-[N-(pyridile-3-carbonyl)amino]ethyl (i), β-(n-tosylamino)ethyl (j).

Synthesized thiol ethers represent viscous oils of light yellow or solid substances with "expanded" melting temperatures, well soluble in the majority of organic substances. Some of them were purified by distillation in vacuum or by recrystallization out of suitable solvents; S-lkyl ethers (indolyl-3) of thioacetic acid LXIX were iolated from reactions mixture and purified by column chromatography on silica gel. Physical-chemical characteristics of obtained thiol ethers LXIXa-j are presented in Table 3.14.

The structures of synthesized compounds were confirmed by data of IR- and PMR-spectroscopies and also by data of elementary analysis. In IR-spectra of thiol ethers LXIXa-j there are intensive absorption maximums in the region 1685-1660 relating to stretching vibrations of carbonyl group of thiocarboxyl fragment [167, 225].

3.2.2. Synthesis of Five-Membered N-Containing Heterocycles Containing β-Alkylthioethyl Fragments

Some of synthesized by us functional derivatives of β-alkyl-thiopropionic acids – hydrochlorides of methyl iminoethers LXIII, N-acylhydrozones of carbonyl compounds LXVIa-c, e, f, N,N'-diacylhydrazines LXVIIa-c, e and 1-acyl-4-R-thiosemicarbazides LXVIIIa-c – later were used as synthons under reception of nitrogen-containing five-membered hetero-cycle compounds with 2 or 3 hetero-atoms. Compounds of such type represent special interest as stabilizers and additives for hydrocarbon fuels, lubricating oils and polymer materials, inhibitors of corrosion and bio-destruction of organic materials, herbicides and fungicides.

Analysis of literature data [174] shows that it is advisable to use reactive salts of iminoethers for reception of mentioned hetero-cycles, in particular dihydroazoles, benzazoles, oxadiazoles and 1,3,4-thiadiazoles on the base of derivatives of carbonic acids. Reactions with participation of iminoethers salts proceed at relatively low temperatures, without catalysts, reactions times are lower and the yield of heterocyclic compounds is higher that under the use of other functional derivatives of carbonic acids (nitriles, amides, esters, chloroanhydrides, anhydrides, etc.).

Table 3.14. Constants, yield and data of elementary analysis of S-alkyl ethers of thiocarbonic acids RC(=O)SAlk

Compound	R	Yield, %	$T_{boil.}$, °C (millimeters of mercury) or $T_{melt.}$, °C (solvent for crystallization)	n_D^{20}	d_4^{20}, g/cm^3	S, mass %
LXIXa	n-C_5H_{11}-	84	60-85 (2)	1.4854	1.0945	17.94
LXIXb	n-$C_{11}H_{23}$-	75	100-122 (1)	1.4960	1.0684	11.52
LXIXc	n-$C_{17}H_{35}$-	74	142-161 (2)	1.5008	1.0982	9.46
LXIXd	Ph-	86	118-139 (2)	1.5046	1.0235	17.34
LXIXe	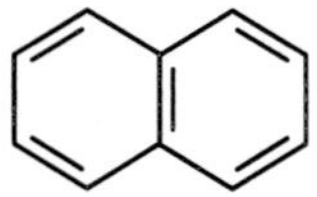	83	44-48 (hexane)	-	-	13.05
LXIXf	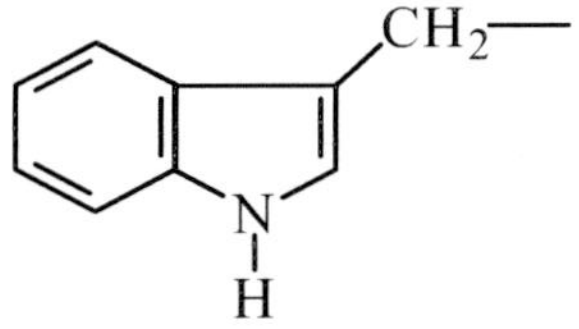	81	oil*)	1.5174	1.0931	12.88
LXIXg	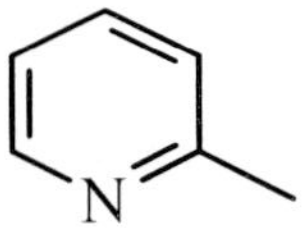	88	39-43 (ethanol-water,1:1)	-	-	17.15
LXIXh	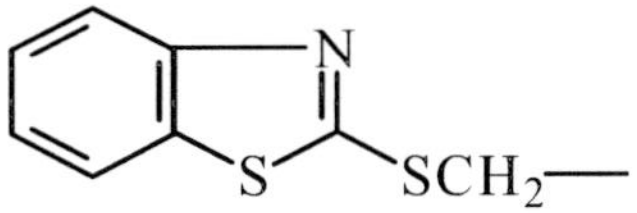	80	60-65 (heptane)	-	-	33.26
LXIXi		84	52-57 (chloroform)	-	-	12.62
LXIXj	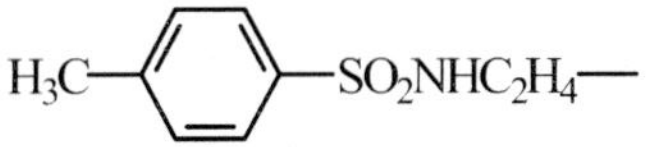	78	84-90 (ethanol-water, 2:1)	-	-	20.43

Note: *) The compounds were purified by column chromatography on silica gel.

As a result of condensation of mixture of iminoethers hydrochlorides LXIII with ethylenediamines, N-monosubstituted ethylendiamines, monoethanolamines and 1-aminopropanol-2 the mixtures of 1-R-2-(β-alkylthioethyl)-Δ^2-imidazolines LXXa-f and 2-(β-alkylthioethyl)-5-R^1-Δ^2-oxazolines LXXIa, b are correspondingly formed.

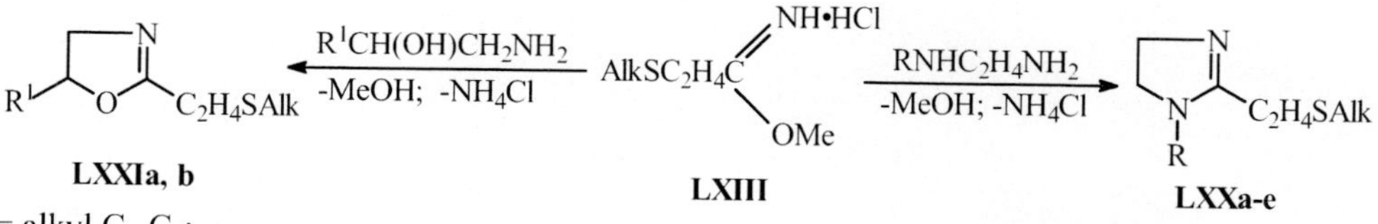

Alk = alkyl C_2-C_5;
70: R = H (a), *n*-C_4H_9 (b), *n*-$C_{12}H_{25}$ (c), $PhCH_2$ (d), 4-HO-3,5-(*tert*-$C_4H_9)_2C_6H_2CH_2$ (e);
71: R^1 = H (a), CH_3 (b).

We found that these condensation reactions should be carried out under boiling of equimolar amounts of reagents in absolute ethanol for 3-4 hours; at these conditions the yield of desired products is 70-84% (Table 3.15).

Table 3.15. Constants, yield and data of elementary analysis of 1-R-2-(β-alkylthioethyl)-Δ^2-imidazolines and 2-(β-alkylthioethyl)-5-R^1-Δ^2-oxazolines

Compound	R	R^1	Yield, %	$T_{mel.}$, °C*)	n_D^{20}	S, mass %
LXXa	H-	-	82	50-56	-	18.13
LXXb	*n*-C_4H_9-	-	75	oil	1.5725	13.24
LXXc	*n*-$C_{12}H_{25}$-	-	70	oil	-	8.91
LXXd	$PhCH_2$-	-	72	oil	1.5836	12.02
LXXe	4-HO-3,5-(*tert*-$C_4H_9)_2C_6H_2CH_2$-	-	78	4451	-	7.88
LXXIa	-	H-	80	oil	1.5758	19.10
LXXIb	-	CH_3-	74	57-63	-	17.65

Note: *) The compounds were purified by column chromatography on silica gel.

It is known [173, 227] that Δ^2-imidazoline-5-ones not substituted in positions 1 and 4 are as a rule compounds non-stable enough which under condensation with aromatic aldehydes give stable 4-arylidene derivatives. For reception of Δ^2-imidazoline-5-ones of such type they use one-stage method consisting in condensation of carbonic acids iminoethers hydrochlorides with hydrochloride of ethyl ether of aminoacetic acid (glycine) and aldehydes in the presence of bases [174].

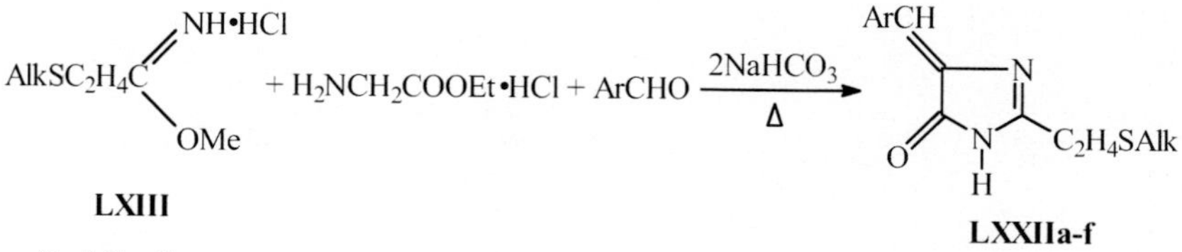

Alk = alkyl C_2-C_5;
Ar = Ph (a), 4-$NO_2C_6H_4$ (b), 2-HOC_6H_4 (c), 4-HO-3,5-(*tert*-$C_4H_9)_2C_6H_2$ (d), 5-nitrofuryl-2 (e), pyridile-3 (f).

We synthesized 2-(β-alkylthioethyl)-4-arylidene-Δ^2-imidazoline-5-ones LXXIIa-f with good yield by boiling (2-3.5 hours) of equimolar amounts of mentioned compounds in dioxane in the presence of sodium bicarbonate (Table 3.16).

Table 3.16. Constants, yield and data of elementary analysis of 2-(β-alkylthioethyl)-4-(Ar-methyl)-Δ^2-imidazoline-5-ones

Compound	Ar	Yield, %	$T_{melt.}$, °C (solvent for crystallization)	S, mass %
LXXIIa	Ph	72	122-128 (benzyl-isooctane, 3:1)	10.84
LXXIIb	4-$NO_2C_6H_4$	76	140-145 (ethanol-water, 2:1)	9.45
LXXIIc	2-HOC_6H_4	68	130-136 (dioxane-water, 1:1)	11.00
LXXIId	4-HO-3,5-(*tert*-$C_4H_9)_2C_6H_2$	62	104-109 (toluene)	7.60
LXXIIe	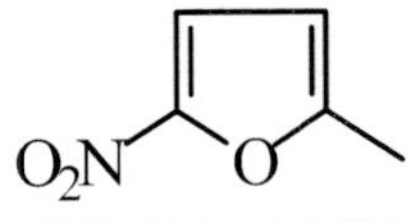	70	138-145 (dioxane-water, 2:1)	9.78
LXXIIf	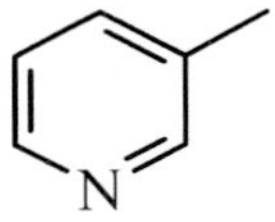	74	100-106 (ethanol-water,1:1)	11.57

The mixtures of 2-(β-alkylthioetyhl)-5-R-benzimidazoles LXXIIIa, b and 2-(β-alkylthioethyl)-5-R^1-6-R^2-benzoxazoles LXXIVa-c were synthesized with high yield by analogous technique in ethanol solution (3-4 hours) with equimolar amount of hydrochlorides of methyl iminoethers LXIII and 4-R-1,2-diaminobenzenes (or 4-R^1-5-R^2-2-aminophenols) (Table 3.17) (method A).

Table 3.17. Constants, yield and data of elementary analysis of 2-(β-alkylthioethyl)-5-R-benzimidazoles and 2-(β-alkylthioethyl)-R^1-6-R^2-benzoxazoles

Compound	R	R^1	R^2	Yield, % (preparative method)	$T_{melt.}$, °C (solvent for crystallization)	S, mass %
LXXIIIa	H-	-	-	85 (A), 46 (B)	112-118 (ethanol-water, 1:1)	13.86
LXXIIIb	CH_3-	-	-	81 (A), 51 (B)	130-136 (ethanol-water, 1:1)	12.94
LXXIVa	-	H-	H-	82 (A)	64-71 (benzene-hexane, 2:1)	14.31
LXXIVb	-	H-	NO_2-	75 (A)	132-137 (dioxane-water, 2:1)	11.74
LXXIVc	-	H_2NSO_2-	H-	73 (A)	103-111 (ethanol-water, 1:1)	21.21

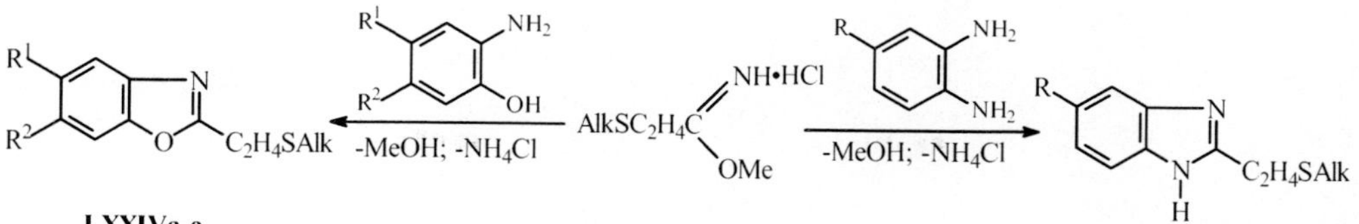

LXXIVa-c LXIII LXXIIIa, b

Alk = alkyl C_2-C_5;
73: R = H (a), CH_3 (b);
74: $R^1 = R^2$ = H (a); R^1 = H, $R^2 = NO_2$ (b); $R^1 = SO_2NH_2$, R^2 = H (c)

Benzimidazoles LXXIIIa, b were also obtained with the yield 46-51% by Phillips' method [173, 228] by boiling (5-6 hours) of mixture of nitriles of β-alkylthiopropionic acids LXI and *o*-phenyldiamine or 4-methyl-*o*-phenylendiamine in ethanol in the presence of 4n hydrochloric acid (method B).

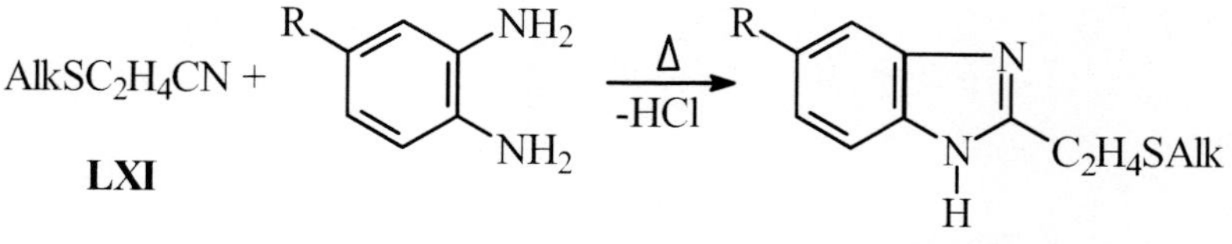

LXI LXXIIIa, b

Derivatives of 1,3,4-oxadiazole are widely used in pharmaceutics, in production of dyes, in colour photography and also as semi-products for reception of thermostable polymer materials [178, 229]. At the same time analysis of literature data showed that before we started our work there was the only one work [218] on the derivatives of 1,3,4-oxadiazole containing β-alkylthioethyl radicals. As it is known [174, 178] condensation of hydrochlorides of iminoethers of carbonic acids with N-acylhydrazines (hydrazides of acids) leads to 2,5-substituted 1,3,4-oxadiazoles.

We used such method for reception of 1,3,4-oxadiazoles including β-alkylthioethyl fragments. With this aim hydrazides of caproic, benzoic, phenylacetic, β-(4-hydroxy-3,5-di-*tert*-butylphenyl)propionic, 5-nitrofuran-2-carbonic, 3-indolylacetic and pyridine-3-carbonic acids were introduced into condensation with hydrochlorides of methyl iminioethers LXIII (method A). Reactions were carried out by boiling of mixture of iminoethers hydrochlorides LXIII with hydrazide of corresponding acid in ethanol for 4-5 hours. As a result of a number of experiments we established that the best yield (73-89%, Table 3.18) of 2-(β-alkylthioethyl)-5-R-1,3,4-oxadiazoles LXXVa-g was obtained at molar ratio hydrochloride LXIII : hydrazide of acid equal to 1.25 : 1.

For reception of 1,3,4-oxadiazoles LXXVb, c, h, i we used cyclodehydration of corresponding N-(β-alkylthioethyl)-N'-acylhydrazines LXVIIa-c, e under the action of phosphorus chloro-oxide [178, 187] at short-term heating however significant resinification of reaction mixture was observed and in this connection the desired products were obtained with the yield 42-50% (method B).

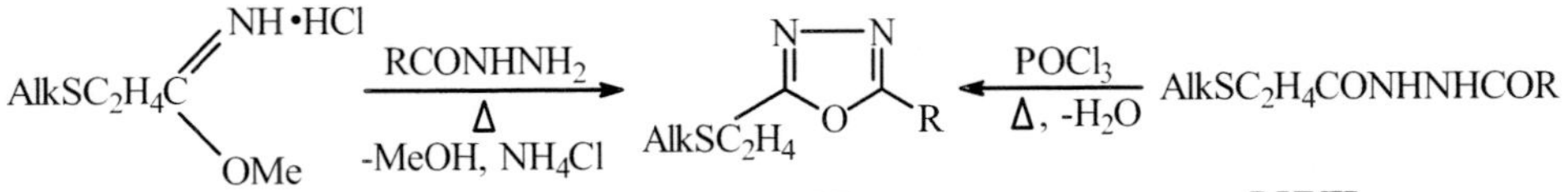

LXIII **LXXV** **LXVIIa-c, e**

Alk = alkyl C_2-C_5;

75: R = n-C_5H_{11} (a), Ph (b), $PhCH_2$ (c), 4-HO-3,5-(*tert*-$C_4H_9)_2C_6H_2CH_2CH_2$ (d), 5-nitrofuryl-2 (e), indolyl-3-methyl (f), pyridile-3 (g), 4-$NO_2C_6H_4$ (h), $AlkSCH_2$ (i).

In accordance with literature data [222, 230-232] N-acylhydrazones of carbonyl compounds containing reactive C=N-groups are suitable synthons under synthesis of di- and three-substituted derivatives of tiazolidine-4-one. During last years the compounds of such type attract essential attention of researches, since in their raw there are low-toxic substances with high and various biologic activity including high-effective antibiotics [230], radioprotectors [233], pesticides [48, 191]. Furthermore derivatives of thiazolidines represent the interest as substances with wide spectrum of applied properties – corrosion inhibitors, vulcanization accelerators, additives of various purposes to lubricating oils, hydrocarbonic fuels, polyolefines and rubbers [72, 234].

Table 3.18. Constants, yield and data of elementary analysis of 2-(β-alkylthioethyl)-5-R-1,3,4-oxadiazoles

Compound	R	Yield, % (preparative method)	$T_{melt.}$, °C (solvent for crystallization)	S, mass %
LXXVa	n-C_5H_{11}-	84 (A)	oil*⁾ (n_D^{20} 1,5592)	13.04
LXXVb	Ph-	88 (A), 44 (B)	124-129 (ethanol-water, 2:1)	12.28
LXXVc	$PhCH_2$-	79 (A), 42 (B)	74-82 (toluene-heptane, 3:1)	12.04
LXXVd	4-OH-3,5-(t-$C_4H_9)_2C_6H_2CH_2CH_2$-	73 (A)	oil*⁾	7.88
LXXVe	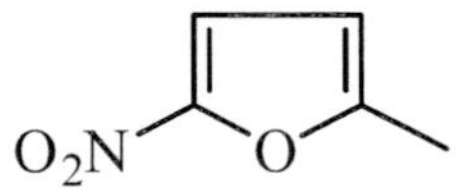	75 (A)	138-145 (DMFA-water, 3:1)	11.25
LXXVf	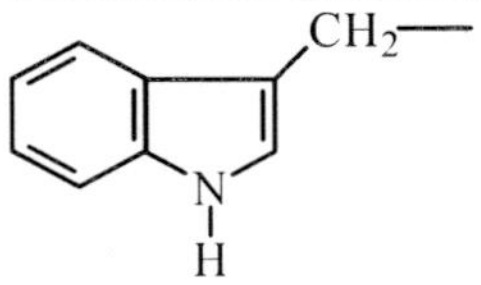	81 (A)	oil*⁾	10.37
LXXVg	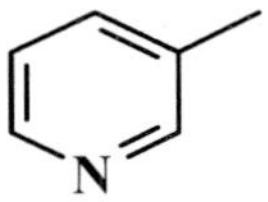	89 (A)	127-134 (ethanol-water, 2:1)	13.04
LXXVh	4-$NO_2C_6H_4$-	50 (B)	140-147 (propanol-2)	10.48
LXXVi	$AlkSCH_2$-	45 (B)	oil*⁾ (n_D^{20} 1,5508)	23.44

Note: *⁾ The compounds were purified by column chromatography on silica gel.

In our work as a result of condensation of N-(β-alkylthiopropionyl)hydrazones LXVIa-f with thioglycolic acid the 2-R-2-R'-3-[N-(β-alkylthiopropionyl)amino]thiazolidine-4-ones LXXVIa-f were synthesized. We established that their best yields (Table 3.19) were obtained under reagents boiling (2 mole excess of thioglycolic acid) in toluene or in DMFA (in the case of hydrazone LXVId) for 8-12 hours in the presence of 5-10 mass % of zinc chloride. In the absence of this catalyst the yield of desired heterocycles is 35-42%.

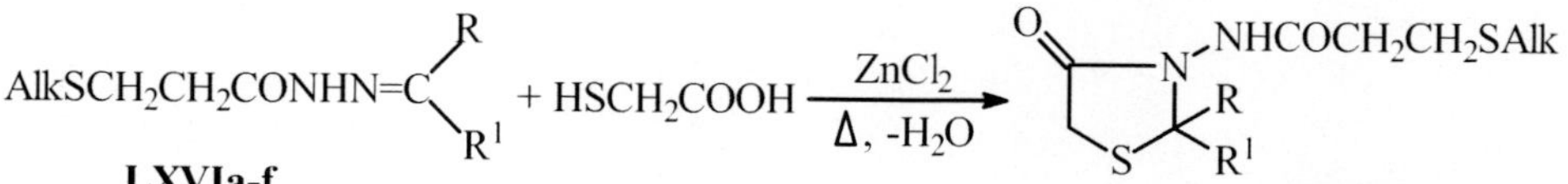

LXVIa-f **LXXVIa-f**

Alk = alkyl C_2-C_5;

R = *iso*-C_3H_7, R^1 = H (a); R = Ph, R^1 = H (b); R = 4-HO-3,5-(*tert*-C_4H_9)$_2$$C_6H_2$, R^1 = H (c); R = 5-nitrofuryl-2, R^1 = H (d); R = R^1 = CH_3 (e); R = Ph, R^1 = CH_3 (f).

We should note that ketones derivatives for example LXVIe, f are transformed into 2,2-disubstituted thiazolidine-4-ones LXXVIe, f more hard and with lower yield than corresponding derivative of aldehydes. On the base of this fact we may conclude that spatial impediments created by two substituents at carbon atom of nitrogenmethine group create hindrances of molecule cyclization.

Table 3.19. Constants, yield and data of elementary analysis of 2-R-2-R'-3-[N-(β-alkylthiopropionyl)amino]thiazolidine-4-ones

Compound	R	R^1	Yield, %	$T_{melt.}$, °C (solvent for crystallization)	S, mass %
LXXVIa	*iso*-C_3H_7-	H-	80	oil*) (n_D^{20} 1,5885)	21.95
LXXVIb	Ph-	H-	82	94-101 (benzene-heptane, 4 : 1)	19.82
LXXVIc	4-HO-3,5-(*tert*-C_4H_9)$_2$$C_6H_2$-	H-	78	52-57 (petroleum-ether)	14.03
LXXVId	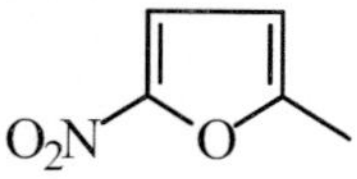	H-	84	142-146,5 (ethanol)	17.72
LXXVIe	CH_3-	CH_3-	72	oil*) (n_D^{20} 1,5842)	23.85
LXXVIf	Ph-	CH_3-	71	60-66 (benzene-heptane, 3 : 1)	18.94

Note: *) The compounds were purified by column chromatography on silica gel.

As it is known [220, 221, 226, 227, 235] 1-acylthiosemicarbazides are suitable initial substances in synthesis of thioderivatives 1H-1,2,4-triazole and amino-derivatives of 1,3,4-oxadiazole. Investigation of heterocyclization of 1-(β-alkylthiopropionyl)-4-R-thiosemicarbazides LXVIIIa-c showed that products character depended on conditions of reaction carrying out. It was established that in alkaline medium compounds LXVIIIa-c were cyclisized into 3-(β-alkylthiopropionyl)-4-R-1H-1,2,4-triazoline-5-thiones LXXVIIa-c. This cyclization should be carried out under boiling (1-1.5 hours) in 5% solution of sodium hydroxide and as intermediates Na-derivatives of 5-mercaptano-4H-1,2,4-triazoles are formed which at 0°C are transformed into corresponding 1,2,4-triazoline-5-thiones LXXVIIa-c with high yield when are acidified by hydrochloric acid (Table 3.19).

At the same time 1-acylthiosemicarbazides LXVIIIa-c when treated by alcohol solution of iodine in alkaline medium are transformed into 2-(R-amino)-5-(β-alkylthioethyl)-1,3,4-oxadiazoles LXXVIIIa-c however their yield does not exceed 55-60%. The rise of reaction time and temperature and also substitution of iodine by more active bromine did not lead to noticeable increase of compounds LXXVIIIa-c yield.

The structures of synthesized compounds LXVIIa-c were confirmed by chemical transformations characteristic for such heterocycles. For example, under alkylation of these compounds by CH_3I, benzylchloride or chloroacetic acid in ethanol in the presence of 5% alkali solution (boiling for 3-5 hours) the 3-(β-alkylthioethyl)-4-R-5-(R^1-amino)-4H-1,2,4-triazoles LXXIXa-g with the yield 72-85% were synthesized (Table 3.20).

Table 3.20. Constants, yield and data of elementary analysis of 3-(β-alkylthioethyl)-4-R-1H-1,2,4-triazoline-5-thiones LXXVIIa-c, 2-(R-amino)-5-(β-alkylthioethyl)-1,3,4-oxadiazoles LXXVIIIa-c and 3-(β-alkylthioethyl)-4-R-5-(R^1-amino)-4H-1,2,4-triazoles LXXIXa-g

Compound	R	R^1	Yield, %	$T_{melt.}$, °C (solvent for crystallization)	S, mass %
LXXVIIa	CH_2=$CHCH_2$-	-	82	72-78*)	25.50
LXXVIIb	Ph-	-	86	128-134 (ethanol-water, 1 : 1)	22.16
LXXVIIc	4-ClC_6H_4-	-	80	140-144 (ethanol-water, 1 : 1)	20.34
LXXVIIIa	CH_2=$CHCH_2$-	-	55	60-64*)	14.10
LXXVIIIb	Ph-	-	60	148-152 (propanol-2)	11.71
LXXVIIIc	4-ClC_6H_4-	-	58	120-125 (benzene-heptane, 3 : 1)	10.42
LXXIXa	CH_2=$CHCH_2$-	CH_3-	77	масло*) (n_D^{20} 1,5792)	24.13
LXXIXb	CH_2=$CHCH_2$-	PhCH_2-	75	48-53 (petroleum-ether)	19.02
LXXIXc	CH_2=$CHCH_2$-	HOOCCH_2-	85	158-164 (propanol-2-water, 2 : 1)	20.34
LXXIXd	Ph-	CH_3-	81	101-105 (benzene-heptane, 2 : 1)	20.88
LXXIXe	Ph-	HOOCCH_2-	80	148-152 (ethanol)	19.06
LXXIXf	4-ClC_6H_4-	CH_3-	76	82-87 (benzene)	19.58
LXXIXg	4-ClC_6H_4-	HOOCCH_2-	83	132-137 (propanol-2-water, 2 : 1)	16.42

Note: *) The compounds were purified by column chromatography on silica gel.

Physical-chemical characteristics of synthesized heterocyclic compounds LXX-LXXIX are presented in Tables 3.15-3.20. Their structures were confirmed by data of IR- and PMR-spectroscopies and also be data of elementary analysis.

In IR spectra of mentioned compounds stretching vibrations of C–S bonds are characterized by absorption bands of variable intensity in the interval 1210-1160cm^{-1} [167]. In spectra of Δ^2-imidazolines LXXa-f there are strong absorption bands in the region 1625-1610cm^{-1} characteristic for stretching vibrations of group C=N in dihydroazoles [181, 182]. Stretching and deformation vibrations of methylene group of imidazoline cycle are presented by absorption bands of variable intensity in the regions 2990-2980, 2915-2875, 1450-1440 ("cutter" vibrations of CH_2), 1255-1240 and 1015-1000cm^{-1} accordingly [181, 182]. In spectra of Δ^2-oxazolines LXXIa, b absorption bands of group C=N are observed at 1665-1650cm^{-1} that is characteristic for heterocycles of this type [183]. The presence of oxazoline cycle is confirmed also by absorption maximums at 1275-1265 and 1015-1010cm^{-1} relating to stretching vibrations of fragment =C–O–C in Δ^2-oxazolines [184].

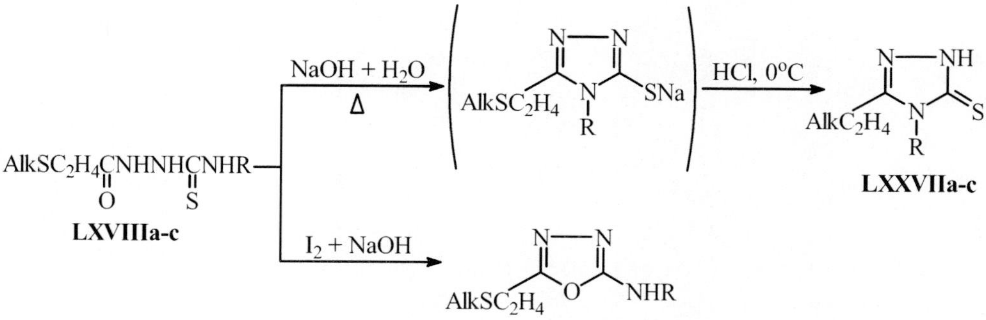

Alk = alkyl C_2-C_5; R = CH_2=$CHCH_2$ (a), Ph (b), 4-$C1C_6H_4$ (c).

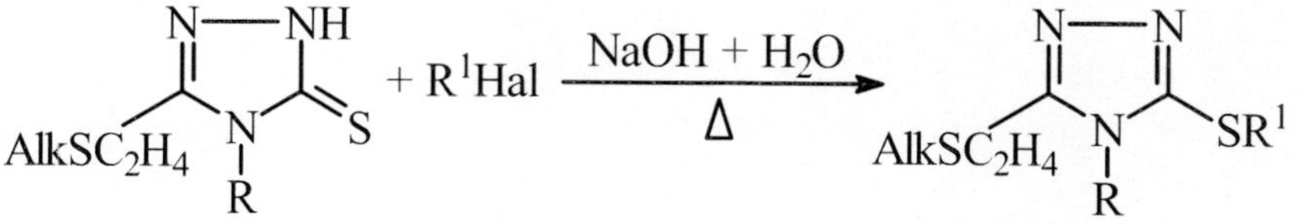

Alk = alkyl C_2-C_5; R = CH_2=$CHCH_2$ (LXXVIa, LXXVIIIa-c), Ph (LXXVIb, LXXVIIId, e), 4-ClC_6H_4 (LXXVIc, LXXVIIIf, g); R^1 = CH_3 (LXXVIIIa, d, f), $PhCH_2$ (LXXVIIIb), $HOOCCH_2$ (LXXVIIIc, e, g); Hal = I,C1.

In spectra of Δ^2-imidazolidine-5-ones LXXIIa-f two intensive absorption bands are revealed at 1610-1605 and 1690-1680cm^{-1} characterising accordingly the stretching vibrations of C=N group and carbonyl group in derivatives of dihydroazoles [183, 236].

In spectra of benzazoles LXXIIIa, b and LXXIVa-c the intensive absorption bands are observed in the region 1640-1625cm^{-1} relating to vibrations of group C=N in 2-substituted benazoles [183, 237]. In spectra of benzimidazoles LXXIIIa, b there are also absorption bands of variable intensity characteristic for stretching (1550, 1485-1480, 1465-1460 and 1330cm^{-1}), "breathing" (1065cm^{-1}) and plane deformation vibrations of benzimidazole cycle [237]. The structure of benzoxazoles LXXIVa-c is confirmed by the presence of intensive absorption bands at 1590-1580 and 935-930cm^{-1} peculiar to benzoxazole cycle and absorption maximums at 1255-1250 and 1020-1010cm^{-1} relating to asymmetric and symmetric vibrations of fragment =C–O–C= of benzoxazoles [183, 238].

In spectra of synthesized 1,3,4-oxadiazoles LXXIVa-i and LXXVIIIa-c in the region of stretching vibrations of C=N groups there absorption bands of variable intensity at 1670-1630cm^{-1} and absorption maximums intensive enough in the interval 1615-1595 and 1490-1470cm^{-1}. the last group of bans is characteristic for stretching vibrations of 1,3,4-oxadiazole cycle [239, 240]. The presence of this cycle is confirmed also by absorption bands at 1245-1225 and 1045-1020cm^{-1} relating to stretching vibrations of fragment =C–O–C= in 1,3,4-oxadiazoles [188] and by absorption maximum near 970cm^{-1} connected with "breathing" vibrations of oxadiazole cycle [241].

In IR spectra of thiazolodine-4-ones LXXVIa-f the absorption bands of variable intensity characterizing thiozolidine cycle were fixed [183]: 1470-1450, 1445-1425 ("cutter" vibrations of CH_2), 1290-1280, 1085-1070 (stretching vibrations of cycle), 1180-1165 and 1010-965cm^{-1}. Intensive absorption maximums in the region 1755-1740cm^{-1} characterize vibrations of carbonyl group in thiazolidinones [183, 231, 242, 243], and in the region 1670-1655cm^{-1} absorption maximums characterize vibrations of "amide I" type of N-acyl group [167, 243]. The last vibrations are shifted to low-frequency region in comparison with stretching vibrations of carbonyl group (1695-1675cm^{-1}) in initial N-acylhydrazones LXVIa-f.

The structure of 3-subtituted 4-R-1H-1,2,4-triazoline-5-thiones LXXVIIa-c and not of tautomeric to them 3-substituted 5-mercaptano-4-R-4H-1,2,4-triazoles was assigned to products of cyclization of 1-acylthiosemicarbazides LXVIIIa-c in alkaline medium on the base of spectral data. The correctness of such interpretation is confirmed by the absence in IR spectra of absorption in the regions 2600-2500cm^{-1} ($\nu_{S\text{-}H}$) and 900-850cm^{-1} (planar $\nu_{C\text{-}SH}$). The absence of mentioned bands correlate with the presence of intensive absorption band at 1530-1520cm^{-1} relating to vibrations of thioamide group [224] and also of two absorption maximums at 1350-1335 and 1320-1305cm^{-1} characteristic for vibrations of group C=S in 1,2,4-triazoline-5-thiones [186, 244]. Absorption maximums at 1440-1430, 1245-1235 and 1155-1140cm^{-1} relate to vibrations of 1,2,4-triazoline cycle [245].

In spectra of compounds LXXIXa-g the number of absorption bands is observed connected with vibrations of 1,2,4-triazoline cycle [49, 50, 145]: 1475-1460, 1390-1380, 1290-1275 and 1030-1015cm^{-1}.

The presence of absorption bands conditioned by fragment of spatially-hindered phenol is characteristic for IR spectra of compounds LXXf, LXXIId, LXXIVd and LXXVIc: narrow band at 3655-3640cm^{-1} (screened hydroxyl group [205]), two bands of average intensity at 1260-1230cm^{-1} (stretching vibrations of C–O in screened phenols [245]) and two groups of bands in the regions 880-870 and 830-820cm^{-1} (intra-planar deformation vibrations of C–H in aromatic cycle).

In PMR spectra of synthesized compounds in the region of resonance of protons of alkyl fragments several groups of signals are observed differing in values of chemical shifts. Signals of protons of methyl groups represent doublets an triplets at 0.84-1.12m.p.. Complex multiplets in the interval 1.18-1.52m.p. should be assigned to protons of methylene units, in the interval 1.70-1.84m.p. – to protons of methane groups, and signals shifted to weak field (2.54-3.50m.p.) – to protons of CH_2– and CH-groups bonded with sulfur atom and also to protons of CH_2-groups bonded with heterocycle rings.

In PMR spectra of Δ^2-imidazolines LXXa-f the signals of 4 protons of cycle are revealed as non-symmetric multiplet in the region 4.10-4.42m.p. (spin system AA'BB') [175]. In Δ^2-oxazolines LXXIa, b spectra signals of cycle protons are observed as multiplets in the interval 3.95-4.30m.p.. Signals of arilidene proton in spectra of Δ^2-imidazoline-5-ones LXXIIa-f are in the shape of singlets posed in weak filed at 7.85-8.25m.p. that is characteristic for compounds of such type. Signals of protons of NH-groups of benzimidazole cycles in spectra of compounds LXXIIIa, b are presented by broadened singlets in the interval 8.12-8.24m.p.. Protons of benzene nucleus in spectra of benzazoles LXXIIIa, b and LXXIVa-c give complex multiplets at 7.30-7.86 in weak filed.

In spectra of thiazolidine-4-ones LXXVIa-f signals of methylene protons in position 5 represent typical AB-system of two non-symmetric doublets at 3.92-4.14 and 4.24-4.48m.p. [167]. Singlets in the region 3.32-3.65m.p. characterize the proton in position 2 in 2-substituted thiazolidine-4-ones LXXVIa-d. Signals of NH-groups of N-acyl fragments are observed as broadened singlets at 10.35-11.05m.p..

In spectra of 1,2,4-triazoline-5-thiones LXXVIIa-c signals of NH-protons are characterized by broadened singlets at 10.35-11.05m.p..

For some of synthesized heterocyclic compounds the contents of individual components in reaction products mixture were determined with the help of chromatography-mass spectrometry (Table 3.21).

In presented Section we described reception of various five-membered nitrogen-containing heterocycles (Δ^2-imidazoline, Δ^2-oxazoline, benzimidazoles, benzoxazoles, 1,3,4-oxadiazoles, 1,2,4-triazoles) including β-alkylthioethyl fragments on the base of functional derivatives of β-alkylthiopropionic acids. We also carried out the synthesis of such compounds by reactions of free radical addition of C_2–C_5 mercaptans to N-vinyl derivatives of nitrogen-containing heterocycles. These reactions are known long ago and are well investigated [36, 79, 165], some of them underlie preparative methods of sulfides synthesis.

Addition of thiols to C=C may be realized (in dependence on conditions) via nucleophilic, electrophilic or radical mechanisms. Nucleophilic addition of thiols to ethylene bond as a rule occurs in alkaline medium or in the presence of basic catalysts and is facilitated in the presence of electron acceptor groups at double bond (Section 3.2.1). Electrophilic addition proceeds via the stage of carbonic cation formation in acetic medium including formation under the action of such catalysts as HCl, $AlCl_3$, SO_2Cl_2, etc.

Table 3.21. Individual composition of mixtures synthesized nitrogen-containing heterocyclic compounds (in accordance with chromatography-mass spectrometry data)

Alk	The content of component in compounds mixture, %			
	LXIXa	LXXIa	LXXVc	LXXVIe
C_2H_5	3.5	2.8	3.2	3.3
n-C_3H_7	20.9	21.6	21.7	21.8
iso-C_3H_7	41.0	40.8	41.6	41.2
n-C_4H_9	6.0	5.6	5.8	5.5
isoo-C_4H_9	21.2	22.7	20.9	22.1
n-C_5H_{11}	1.9	1.6	1.8	1.6
iso-C_5H_{11}	5.5	4.9	5.0	4.5

Radical addition of thiols to double bond is initiated by peroxides or by UV-irradiation and is often accompanied by isomerization of initial unsaturates:

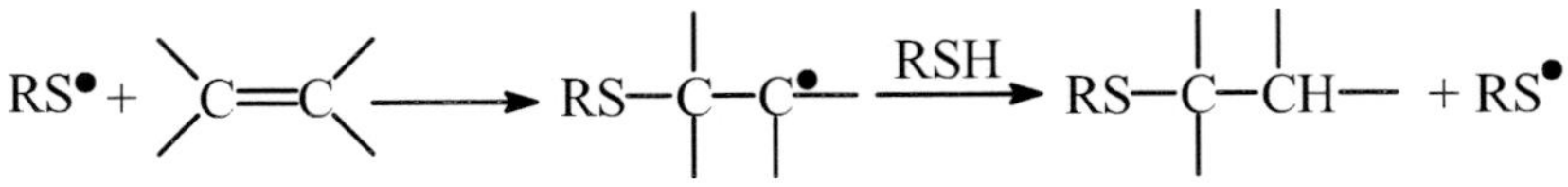

However at present the data on radical addition of alkanethioles to N-vinyl derivatives of nitrogen-containing heterocycles are limited [165, 246, 247].

We studied the interaction of C_2–C_5 mercaptans mixture with 1-vinyl-2-phenylpyrrole, 1-vinyl- and 1-vinyl-7-methyl-4,5,6,7-tetrahydroindole, 9-vinylcarbazole, 1-vinylimidazole and 1-vinyl-2-methylbenzimidazole in the presence and in the absence of initiator. When varying conditions of experiments we established that addition reaction proceeded easily for 18-22 hours under heating of equimolar amounts of reagents in sealed off ampoules up to 70-80°C in the presence of 0.5-1.0 mass % of dinitrile of *iso*-butyric acid. The yield of desired N-β-alkylthioethyl derivatives of heterocycles LXXXa-f is 85-93%, and when initiator is absent – 20-30%.

$$\text{AlkSH} + \text{Het—CH=CH}_2 \xrightarrow[70\text{-}80^{\circ}\text{C}]{\text{Initiator}} \text{HetCH}_2\text{CH}_2\text{SAlk}$$

LXXXa-f

Alk = alkyl C_2-C_5; Het = 2- phenylpyrrolyl-1 (a), 4,5,6,7-tetrahydroindolyl-1 (b), 7-methyl-4,5,6,7-tetrahydroindolyl-1 (c), carbazolyl-9 (d), imidazolyl-1 (e), 2-methylbenzimidazolyl-1 (f).

The particularity of reaction is its high regio-selectivity: in reaction mixture there are no by-products including polymer ones at all at any conversion degree.

Physical-chemical characteristics of synthesized sulfides LXXXa-f are presented in Table 3.22. Structures of these compounds were confirmed by data of IR and PMR-spectra and also by elementary analysis data. In IR-spectra there are absorption bands characteristic for heterocyclic system, bands of stretching vibrations of C–S bonds [167] in the interval 1215-1170cm^{-1} and there is no absorption characterizing stretching vibrations of C=C bond in the region 1660-1650, 1630-1620 and 1560-1545cm^{-1}. In PMR spectra of adducts LXXXa-f the signals of methylene protons of CH_2N group are observed as triplets in the region 3.86-4.18m.p., and signals of methylene and methane groups bonded with sulfur atom – in the shape of complex multiplets in stronger field (3.26-3.45m.p.).

Table 3.22. Constants, yield and data of elementary analysis of N-(β-alkylthioethyl)heterocycles $HetCH_2CH_2SAlk$

Compound	Het	Yield, %	$T_{boil.}$, °C (millimeters of mercury) or $T_{melt.}$, °C (solvent for crystallization)	n_D^{20}	d_4^{20}, g/cm^3	S, mass %
LXXXa	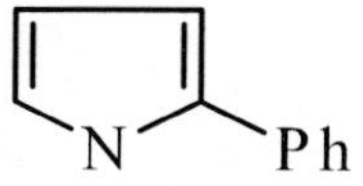	88	148-163 (1)	1.5758	1.0421	12.27
LXXXb	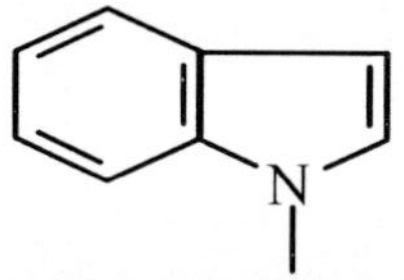	90	125-142 (1)	1.5422	1.0108	13.84
LXXXc	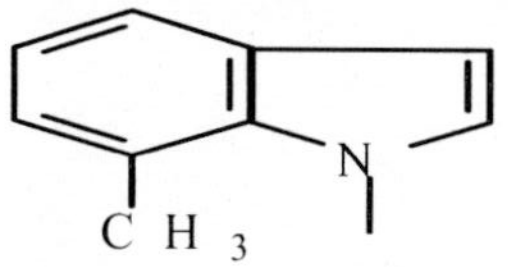	85	132-148 (1)	1.5388	1.0205	13.11
LXXXd	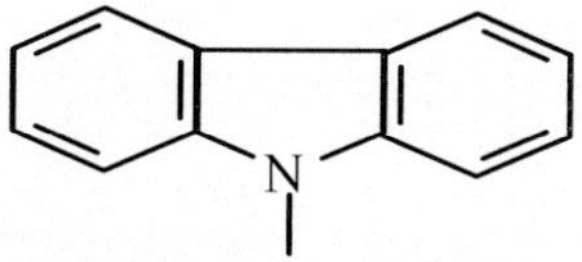	93	122-128 (propanol-2)	-	-	11.40
LXXXe	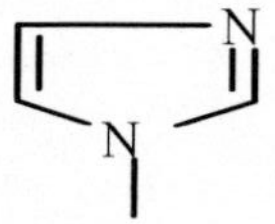	87	124-142 (2)	1.5574	0.9818	18.08
LXXXf	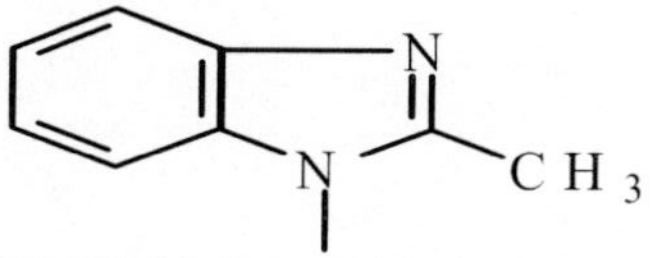	91	102-109 (petroleum-ether)	-	-	12.97

Chapter 4

SYNTHESIS AND TRANSFORMATIONS OF PRODUCTS OF THIOALKYLATION OF MERCAPTANS C_2–C_5

At present all reactions in which substitution of mobile H atoms by alkylthiomethyl (or other alkylthioalkyl groups) occurs under the action of formaldehyde (or other aldehydes) and thiole on compounds with such atoms are nominally related to thioalkylation reactions [31, 248].

$$RSH + CH_2O + R^1H \longrightarrow RSCH_2R^1 + H_2O$$

These reactions are analogous of K. Mannich reaction [33, 249, 250] and are particular case of general reaction of thiomethylation [251].

4.1. SYNTHESIS OF ALKYLTHIOMETHYLAMINES BY THIOMETHYLATION OF AMINES OF VARIOUS STRUCTURES

In some works [248, 252] they consider well-known reaction of aminomethylation of thioles as thioalkylation in which ammonia or primary and secondary amines behave as compounds with mobile protons.

$$RSH + CH_2O + R^1R^2NH \longrightarrow RSCH_2NR^1R^2 + H_2O$$

Thioalkylation of RNH_2 and R^1R^2NH proceeds by the mechanism analogous to K. Mannich reaction [33, 252]: via intermediate formation (in alkaline medium) of aminocarbinol $R^1R^2N–CH_2OH$ or (in acetic medium) carbonium-imine ions $R^1R^2N–CH_2^+ \leftrightarrow R^1R^2N^+ = CH_2$.

In our work we studied thiomethylation of various primary and secondary aliphatic, aromatic, and heterocyclic amines under interaction of mercaptans C_2–C_5 mixture with water solution of formaldehyde. As a result of carrying out of this reaction in ethanol N,N-disubstituted alklthiomethylamines LXXXIa-f are formed (method A).

$$\mathrm{AlkSH + CH_2O + R^1RNH \xrightarrow[-H_2O]{\Delta} AlkSCH_2NRR^1}$$

LXXXIa-f

Alk = alkyl C_2-C_5; R = R^1 = C_2H_5 (a); R = *n*-$C_{18}H_{37}$, R^1 = CH_3 (b); R = R^1 = $PhCH_2$ (c); R = Ph, R^1 = CH_3 (d); R^1RN = piperidino (e), morfolino (f)

When varying conditions in a number of experiments we established that it was advisable to carry out mentioned three-component reaction in accordance with the following technique: equimolar amount of secondary amine is added to the solution of thioles mixture, then equimolar amount of 37% water solution of formaldehyde (formalin) and reaction mixture are boiled for 3-4hours. At these conditions the yield of desired amines LXXXIa-f is 63-82% (Table 4.1).

At analogous conditions on the base of primary amines N-substituted alkylthiomethylamines LXXXIg-n were synthesized with the yield 48-65% (Table 4.1); at that the corresponding N-substituted di(alkylthiomethyl) amines LXXXIIa-h were formed as by-products (method A). the yield of the last ones was from 6-13% (for aromatic and heteroaromatic amines) up to 26-33% (for butyl-, benzyl- and furfurylamines).

$$\mathrm{AlkSH + CH_2O + RNH_2 \xrightarrow[-H_2O]{\Delta} AlkSCH_2NHR + AlkSCH_2N(R)CH_2SAlk}$$

LXXXIg-n **LXXXIIa-h**

Alk = alkyl C_2-C_5; R = *n*-C_4H_9 (LXXXIg, LXXXIIa), $PhCH_2$ (LXXXIh, LXXXIIb), 4-$CH_3C_6H_4$ (LXXXIi, LXXXIIc), 4-HOC_6H_4 (LXXXIj, LXXXIId), 4-HO-3,5-(*tert*-C_4H_9)$_2C_6H_2$ (LXXXIk, LXXXIIe), furfuryl (LXXXIl, LXXXIIf), pyridile-2 (LXXXIm, LXXXIIg), benzthiazolyl-2-thio (LXXXIn, LXXXIIh)

Carrying out of this reaction under boiling for 4-5hours in ethanol and at molar ratio thioles : formaldehyde : primary amine equal to 2 : 2 : 1 leads to tertiary amines LXXXIIa-h with the yield 72-85% (Table 4.2), and even traces of secondary amines LXXXIg-n were not found in reaction mixture.

$$\mathrm{2AlkSH + 2CH_2O + RNH_2 \xrightarrow[-2H_2O]{\Delta} (AlkSCH_2)_2NR}$$

LXXXIIa-h

The synthesized secondary LXXXIg-n and tertiary amines LXXXIa-f, LXXXIIa-h containing alkylthiomethyl groups represent viscous light-yellow or yellow oils or solid substances with "expanded" melting temperatures and are well soluble in ether, acetone, methylenechloride, dioxane, alcohols and aromatic hydrocarbons. Some of these amines were purified by distillation in vacuum or by recrystallization out of suitable solvents; compounds LXXXIk, n and LXXXIIf, h were isolated from reaction mixtures and purified by column chromatography on silica gel. Physical-chemical characteristics of synthesized alklythiomethylamines LXXXIa-n and LXXXIIa-h are presented in Table 4.2.

Table 4.1. Constants, yield and data of elementary analysis of N-mono- and N,N-disubstituted alkylthiomethylamines $AlkSCH_2NRR$

No.	R	R[1]	Yield, % (preparative method)	$T_{boil.}$, °C (millimeters of mercury) or $T_{melt.}$, °C (solvent for crystallization)	n_D^{20}	d_4^{20}, g/cm³	S, mass %
LXXXIa	C_2H_5-	C_2H_5-	78 (A)	50-75 (5)	1.4850	0.9402	19.22
LXXXIb	*n*-$C_{18}H_{37}$-	CH_3-	63 (A)	45-49 (acetone)	-	-	8.24
LXXXIc	$PhCH_2$-	$PhCH_2$-	72 (A)	86-107 (2)	1.4874	0.9581	11.55
LXXXId	Ph-	CH_3-	76 (a)	62-86 (2)	1.4902	0.9524	15.31
LXXXIe	-$(CH_2)_5$-		83 (A)	50-74 (2)	1.4842	0.9460	17.50
LXXXIf	-$(CH_2)_2$-O-$(CH_2)_2$-		80 (A)	54-78 (2)	1.4885	0.9442	17.38
LXXXIg	*n*-C_4H_9-	H-	48 (A)	43-68 (2)	1.4758	0.9512	18.54
LXXXIh	$PhCH_2$-	H-	52 (A)	66-89 (1)	1.4894	0.9492	15.48
LXXXIi	4-$CH_3C_6H_4$-	H-	62 (A)	79-101 (1)	1.4968	0.9612	16.01
LXXXIj	4-HOC_6H_4-	H-	60 (A)	52-57 (ethanol-water, 1:1)	-	-	15.84
LXXXIk	4-HO-3,5-(*t.*-C_4H_9)$_2$-C_6H_2-	H-	58 (A), 66 (B)	oil*⁾	1.5109	-	9.92
LXXXIl	Furyl-2-methyl-	H-	50 (A)	75-97 (2)	1.4918	0.9710	15.45
LXXXIm	Pyridine-3	H-	63 (A), 72 (B)	74-79 (benzene-dioxane, 3:1)	-	-	17.15
LXXXIn	Benzthiazolyl-2-	H-	65 (A), 73 (B)	oil*⁾	1.5322	-	25.82
LXXXIo	4-HO-3,5-(*t.*-C_4H_9)$_2$-C_6H_2-	4-HO-3,5-(*t*-C_4H_9)$_2$-C_6H_2-	74 (B)	54-60*⁾	-	-	5.87
LXXXIp	4-$PhNHC_6H_4$-	H-	70 (B)	108-113 (dioxane-water, 2:1)	-	-	10.94
LXXXIq	2,2,6,6-tetramethylpiperidile	H-	70 (B)	95-115 (4)	1.4952	0.9412	13.04

Note: *⁾ The compounds were purified by column chromatography on silica gel.

Table 4.2. Constants, yield and data of elementary analysis of N-substituted di(alkylthiomethyl)amines $(AlkSCH_2)_2NR$

No.	R	Yield %	$T_{boil.}$, °C (millimeters of mercury) or $T_{melt.}$, °C (solvent for crystallization)	n_D^{20}	d_4^{20}, g/cm^3	S, mass %
LXXXIIa	n-C_4H_9-	75	73-95 (1)	1.4834	0.9682	24.17
LXXXIIb	$PhCH_2$-	78	114-134 (1)	1.4933	0.9638	21.42
LXXXIIc	4-$CH_3C_6H_4$-	80	76-81 (benzene-hexane, 4:1)	-	-	20.97
LXXXIId	4-HOC_6H_4-	72	70-74 (propanol-2)	-	-	21.33
LXXXIIe	4-HO-3,5-(*tert*-C_4H_9)$_2C_6H_2$-	83	49-55 (petroleum-ether)	-	-	14.88
LXXXIIf	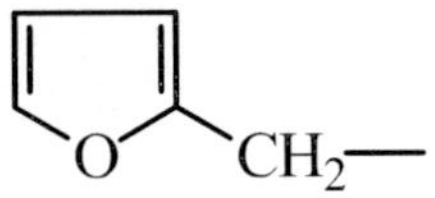	77	oil*)	1.5008	0.9557	24.21
LXXXIIg	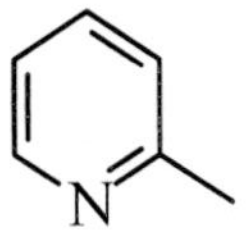	85	40-45 (heptane)	-	-	22.65
LXXXIIh	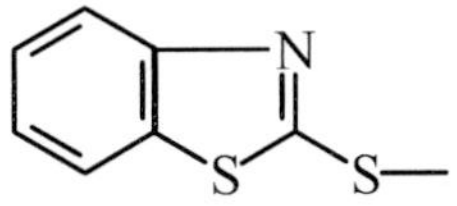	74	oil*)	1.5412	-	32.94

Note: *) The compounds were purified by column chromatography on silica gel.

Structures of synthesized compounds were confirmed by data of IR- and PMR-spectra. In IR-spectra of obtained amines there are absorption bands of variable intensity in the regions 2700-2600, 1435-1420 and 800-690cm^{-1} relating to vibrations of bonds C–S [23]. In spectra of amines LXXXIg-n there are absorption maximums at 3330-3240 and 1655-1640cm^{-1} characteristic correspondingly to stretching and deformation vibrations of NH in secondary amino-groups [167].

In PMR spectra of all synthesized compounds the protons of methylene groups $-SCH_2N<$ are characterized by singlet signals at 3.80-4.08m.p.. In spectra of compounds LXXXIg-n the signals of protons of secondary amino-groups are presented by broadened singlets in the region 5.80-6.25m.p. [167].

4.2. Synthesis of Alkyl(Chloromethyl)Sulfides and Their Derivatives

The known reaction of chloromethylation of thioles (H.Böhme reaction) may be considered as reaction of thioalkylation in which HCl represents compound with mobile hydrogen atom [248, 251, 253].

$$HCl + CH_2O + RSH \longleftrightarrow RSCH_2Cl + H_2O$$

Usually Kh. Byeme reaction is carried out by passing of current of dry HCl through thiole mixture and paraform at –10-15°C in solvent (dichloromethane, benzene) or without it, the yield of desired alkyl- or aryl(chloromethyl)sulfides is 30-65% [253-255]. The mechanism of thioalkylation of HCl is analogous to Mannich reaction since it includes intermediate formation of carbonium- sulfonium ions [248, 253].

In our work for the reception of alkyl(chlorometyhl)sulfides LXXXIII on the base of mixture of natural mercaptans we used modified synthesis method [256] consisting in interaction of mercaptans with paraform and trimethylchlorosilane and excluding application of gaseous HCl, i.e. this method is more convenient for preparative needs in comparison with H.Böhme reaction.

When varying conditions in a number of experiments we found that interaction of mercaptans mixture C_2–C_5 with paraform at the excess of trimethylchlorosilane and 0°C for 3 hours may lead to desired alkyl(chloromethyl)sulfides LXXXIII with the yield 70-74% and also to by-products di(alkylthio)methane LXXXIV with the yield 20-28%.

$$AlkSH + (CH_2O)_n + 2ClSiMe_3 \xrightarrow{0^oC} \underset{\textbf{LXXXIII}}{AlkSCH_2Cl} + Me_3SiOSiMe_3 + HCl$$

$$AlkSCH_2Cl + AlkSH \longrightarrow \underset{\textbf{LXXXIV}}{AlkSCH_2SAlk} + HCl$$

Alk = alkyl C_2-C_5

In accordance with existing at present literature data [253] on high reactivity of chloromethylsulfides in synthesis of various alkyl- and arylthiomethyl derivatives in presented work compounds LXXXIII were used as key substances for reception of N-alkylthiomethyl derivatives of aromatic and heterocycle amines, hydrazides of carbonic acids and also alyklthiomethylthiole derivatives of aliphatic, aromatic and heterocyclic types. So, we showed that interaction of chloromethylsulfides with 4-hydroxy-3,5-di-*tert*-butylaniline, di(4-hydroxy-3,5-di-*tert*-butylphenyl)amine, *n*-aminodiphenylamine, 2-aminopyridine, 2-aminothiazole and 4-amino-2,2,6,6-tetramethylpiperidine in waterless dioxane at 15-20°C (2 hours) with further treatment of reaction mixture by water solution of sodium bicarbonate led to corresponding N-substituted and N,N-disubstituted alkylthiomethylamines LXXXIk, m-q (method B) with high yields (Table 4.1).

$$\mathrm{AlkSCH_2Cl + RR^1NH \xrightarrow[-HCl]{20^oC} AlkSCH_2NRR^1}$$

LXXXIk-q

Alk = alkyl C_2-C_5; R = 4-HO-3,5-(*tert*-$C_4H_9)_2C_6H_2$, R^1 = H (LXXXIk); R = pyridile-2, R^1 = H (LXXXIm); R = thiazolyl-2, R^1 = H (LXXXIn); R = R^1 = 4-HO-3,5-(*tert*-$C_4H_9)_2C_6H_2$ (LXXXIo); R = 2,2,6,6-tetramethylpiperidile-4, R^1 = H (LXXXIp); R = 4-$PhNHC_6H_4$, R^1 = H (LXXXIq)

At analogous conditions N-acyl-N'-alkylthiomethylhydrazines LXXXVa-f were synthesized on the base of hydrazides of 4-nitrobenzoic, 3,4,5-trimethoxy-benzoic, β-(4-hydroxy-3,5-di-*tert*-butylphenyl)propionic and β-(4-hydroxy-3,5-di-*tert*-butylphenylthio)propionic acids and also mixture of hydrazides of alkylthioacetic XI and β-alkylthiopropionic acids LXV (Table 4.3).

$$\mathrm{R\text{-}C(=O)\text{-}NHNH_2 + AlkSCH_2Cl \xrightarrow[-HCl]{20^oC} R\text{-}C(=O)\text{-}NHNH\text{-}CH_2SAlk}$$

LXXXIII **LXXXVa-f**

Alk = alkyl C_2-C_5; R = 4-$NO_2C_6H_4$ (a), 3,4,5-$(CH_3O)_3C_6H_2$ (b), 4-HO-3,5-(*tert*-$C_4H_9)_2C_6H_2CH_2CH_2$ (c), 4-HO-3,5-(*tert*-$C_4H_9)_2C_6H_2SCH_2CH_2$ (d), $AlkSCH_2$ (e), $AlkSCH_2CH_2$ (f).

Table 4.3. Constants, yield and data of elementary analysis of N-acyl-N'-alkylthiomethylhydrazines $RCONHNHCH_2SAlk$

No.	R	Yield, %	$T_{melt.}$, °C (solvent for crystallization)	n_D^{20}	S, mass %
LXXXVa	4-$NO_2C_6H_4$-	82	108-116 (ethanol-water,3:1)	-	11.14
LXXXVb	3,4,5-$(C_2H_5O)_3C_6H_2$-	76	84-91 (benzene)	-	8.42
LXXXVc	4-HO-3,5-(*tert*-$C_4H_9)_2C_6H_4(CH_2)_2$-	78	oil*)	1.5074	7.88
LXXXVd	4-HO-3,5-(*tert*-$C_4H_9)_2C_6H_4S(CH_2)_2$-	72	37-44*)	-	15.52
LXXXVe	$AlkSCH_2$-	84	oil*)	1.4864	25.78
LXXXVf	$AlkSCH_2CH_2$-	80	oil*)	1.4882	24.37

Note: *) The compounds were purified by column chromatography on silica gel.

Table 4.4. Constants, yield and data of elementary analysis of alkylthio(R-thio)methanes $AlkSCH_2SR$

No.	R	Yield, %	$T_{boil.}$, °C (millimeters of mercury) or $T_{melt.}$, °C (solvent for crystallization)	n_D^{20}	d_4^{20}, g/cm^3	S, mass %
LXXXVIa	$PhCH_2$-	79	97-118 (1)	1.4984	0.9705	28.58
LXXXVIb	4-HO-3,5-(трет-C_4H_9)$_2C_6H_2$-	83	137-155 (2)	1.5018	1.0520	19.32
LXXXVIc	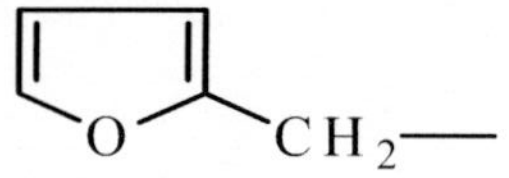	72	112-132 (2)	1.5048	0.9812	29.92
LXXXVId	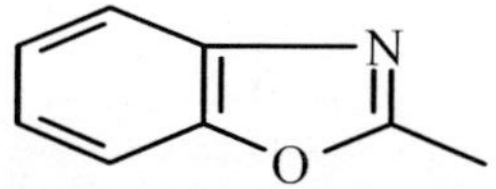	76	136-154 (1)	1.5274	0.9878	26.31
LXXXVIe	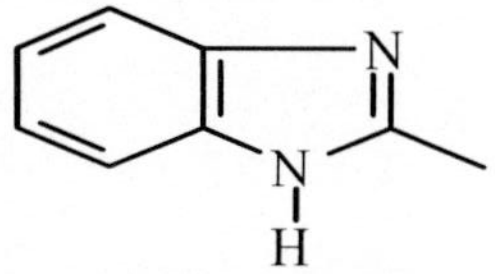	77	108-114 (propanol-2-water, 1:1)	-	-	26.25
LXXXVIf	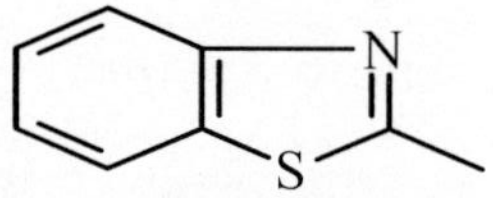	75	152-168 (2)	1.5524	1.1804	36.21
LXXXVIg	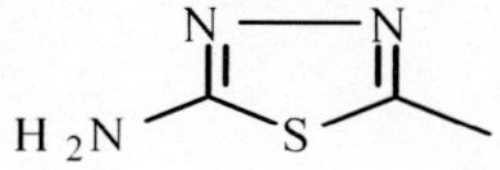	80	172-178 (ethanol)	-	-	41.50
LXXXVIh	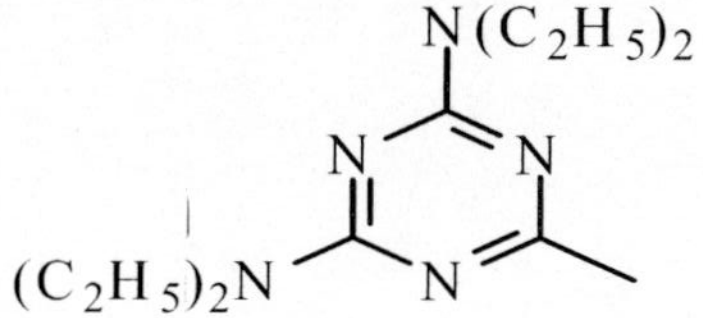	85	72-78 (dioxane-water, 1:2)	-	-	18.33

The corresponding alkylthio(R-thio)methanes LXXXVUa-h with the yield 72-85% were obtained at 0°C in dioxane in the presence of equimolar amount of 30% solution of sodium hydroxide by interaction between chloromethylsulfides LXXXIII with the following mercaptans: benzylmercaptans, 4-hydroxy-3,5-di-*tert*-butylthiophenol, furfurylmercaptane, 2-mercaptobenzoxazole, 2-mercaptobenzimidazole, 2-mercaptobenzthiazole, 2-amino-5-mercapto-1,3,4-thiadiazole and 2-mercapto-4,6-bis(doethylamino)-*sym*-triazine (Table 4.4).

$$\underset{\textbf{LXXXIII}}{AlkSCH_2Cl} + RSH \xrightarrow[NaOH]{0^oC} \underset{\textbf{LXXXVIa-h}}{AlkSCH_2SR}$$

Alk = alkyl C_2-C_5; R = $PhCH_2$ (a), 4-HO-3,5-(*tert*-$C_4H_9)_2C_6H_2$ (b), furfuryl (c), benzoxazolyl-2 (d), benzimidazolyl-2 (e), benzthiazolyl-2 (f), 5-amino-1,3,4-thiadiazolyl-2 (g), 4,6-*bis*(diethylamino)-*sym*-triazinyl-2 (h)

Constants, yield and data of elementary analysis of synthesized alkylthiomethylamines LXXIXk, m-q, alkylthiomethyl hydrazones LXXXIIIa-f and alkylthiomethylsulfides LXXXIVa-h are presented in Tables 4.1, 4.3 and 4.4. structures of these compounds were confirmed by IR and PMR spectra. So, in IR-spectra of synthesized sulfides and *bis*-sulfides there absorption bands of variable intensity in the regions 2700-2600, 1430-1420 and 800-690cm^{-1} relating to vibrations of bonds C-S [167].

In spectra of compounds LXXXIk, m, n, p, q and LXXXVa-f there absorption maximums at 3330-3240 and 1655-1640cm^{-1} characteristic correspondingly for stretching and deformation vibrations of NH in secondary amino-groups [167]. Intensive absorption maximums at 1690-1680cm^{-1} in spectra of N-acylhydrazines LXXXVa-f should be assigned to vibrations of "amide I" type of N-acyl group [209, 211]. In spectra of compounds LXXXIk, p, LXXXVc, d and LXXXVIb the absorption bands conditioned by the fragment of spatially-hindered phenol were also noticed: narrow band at 3655-3640cm^{-1} characteristic for screened phenol hydroxyl [205], two bands of average intensity in the interval 1250-1215cm^{-1} caused by vibrations of C–H bonds in aromatic cycle of screened phenols and two groups of bands in the region 885-870 and 830-815cm^{-1} (intra-planar deformation vibrations of four-substituted benzene cycle).

In PMR spectra of compounds LXXXIk, o, LXXXVc, d and LXXXVIb the signals of protons of hydroxyl groups are revealed in the region 5.10-5.16m.p. in the shape of singlets characteristic for spatially-hindered phenols [205]. Such singlet signals at 7.12-7.28m.p. characterize two magnito-equivalent protons of 4-hydroxy-3,5-di-*tert*-butylphenyl fragments.

In PMR spectra of alkylthiomethylamines LXXXIk, m, n, p, q and alkylthiomethylhydrazines LXXXVa-f the signals of protons of secondary amino-groups are presented as broadened singlets in the region 5.80-6.25m.p. [167]. In spectra of all synthesized compounds singlet signals at 3.80-4.36m.p. characterize protons of methylene groups $-SCH_2N<$ and $-SCH_2S-$.

The contents of individual components in reaction products mixture were determined for some of synthesized compounds with the help of chromatography-mass spectrometry (Table 4.5).

Table 4.5. Individual compositions of mixtures of synthesized alkylthiomethyl derivatives (in accordance with chromatography-mass spectrometry data)

Alk	Content of component in compounds mixture, %					
	LXXXIg	LXXXIk	LXXXIII	LXXXVe	LXXXVIa	LXXXVIc
C_2H_5	2.8	3.0	3.5	2.6	2.5	3.8
n-C_3H_7	20.9	21.0	21.5	20.8	21.7	22.0
iso-C_3H_7	39.5	41.5	42.2	41.7	40.2	38.7
n-C_4H_9	5.7	5.9	5.6	5.5	5.6	5.7
iso-C_4H_9	25.9	22.9	21.3	23.5	24.0	24.2
n-C_5H_{11}	0.9	1.2	1.4	1.3	1.5	1.2
iso-C_5H_{11}	4.3	4.5	4.5	4.6	4.5	4.4

4.3. SYNTHESIS AND CHEMICAL TRANSFORMATION OF γ-KETOSULFIDES

4.3.1. Synthesis of γ-Ketosulfides on the Base of C2–C5 Mercaptans Mixture

One of the most well studied reactions of thioalkyltion is interaction of all ketones (except aromatic ones) containing mobile hydrogen atom in α-position with formaldehyde and mercaptans in alkaline medium leading to γ-ketosulfides (β-alkylthioketones) [31, 86, 133, 257]. It was experimentally established [248, 258] that this reaction in the presence of sodium hydroxides proceeded by two directions (A and B) and the first of them was more likely realized with intermediate formation of hydroxymethylsulfides.

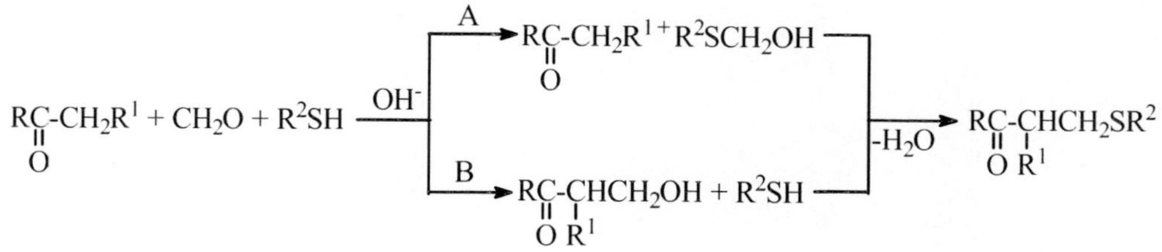

In our work we established on the examples of interaction in water ethanol of mercaptans C_2–C_5 mixture with 35% water solution of formaldehyde, sodium hydroxide and ketones: acetone, butanone, 3,3-dimethylbutanone, acetophenone, 2-acetyl-furan and cyclopentanone that at molar ratio of mentioned reagents 1 : 1 : 1 : 4 and temperature 40-50°C the corresponding mixtures of β-alkylthioketones LXXXVIIa-f were formed with the yield 67-82% for 4-6 hours (Table 4.6).

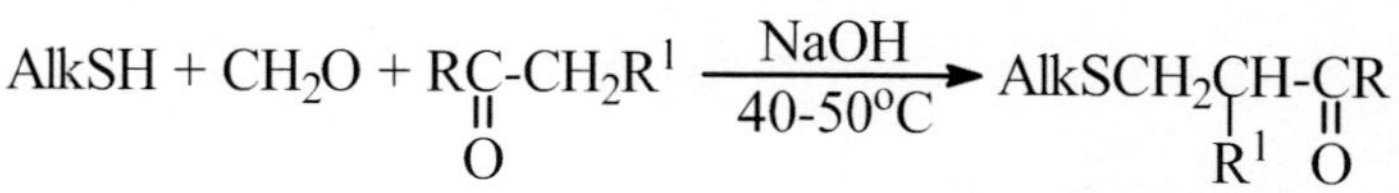

LXXXVIIa-f

Alk = alkyl C_2-C_5; R = CH_3, R^1 = H (a); R = R^1 = CH_3 (b); R = *tert*-C_4H_9, R^1 = H (c); R = Ph, R^1 = H (d); R = furyl-2, R^1 = H (e); R + R^1 = -$(CH_2)_3$- (f)

Change of molar ratio of reagents under interaction of acetone, butanone, 3,3-dimethylbutanone and acetophenone with mercaptans C_2–C_5, formaldehyde solution and alkali at analogous conditions up to 1 : 2 : 2 : 2 gives β,β'-di(alkylthio)ketones LXXXVIIIa-d with the yield 64-73% (Table 4.6).

Table 4.6. Constants, yield and data of elementary analysis of β-alkylthioketones $AlkSCH_2CHR^1COR$ and β,β'-di(alkylthio)ketones $(AlkSCH_2)_2CR^1COR$

No.	R	R^1	Yield %	$T_{boil.}$, °C (millimeters of mercury) or $T_{melt.}$, °C (solvent for crystallization)	n_D^{20}	d_4^{20}, g/cm^3	S, mass %
LXXXVIIa	CH_3-	H-	78	53-78 (3)	1.4708	0.9427	20.14
LXXXVIIb	CH_3-	CH_3-	74	59-83 (3)	1.4735	0.9388	18.68
LXXXVIIc	*tert*-C_4H_9	H-	82	72-95 (2)	1.4810	0.9462	16.21
LXXXVIId	Ph-	H-	67	101-122 (1)	1.5042	0.9718	15.14
LXXXVIIe	furyl-2-	H-	71	99-120 (1)	1.4956	0.9645	15.92
LXXXVIIf	-$(CH_2)_3$-		75	76-90 (1)	1,4782	0.9407	17.40
LXXXVIIIa	CH_3-	H-	73	71-95 (1)	1.5014	0.9514	25.75
LXXXVIIIb	CH_3-	CH_3-	68	92-114 (2)	1.5051	0.9532	21.86
LXXXVIIIc	*tert*-C_4H_9	H-	64	110-130 (2)	1.5082	0.9554	20.21
LXXXVIIId	Ph-	H-	70	52-57 (heptane-methylethylketone, 5:1)	-	-	18.84

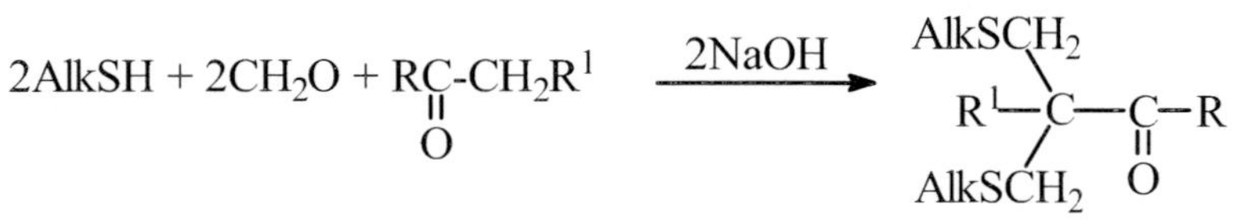

LXXXVIa-d

Alk = alkyl C_2-C_5; R = CH_3, R^1 = H (a); R = R^1 = CH_3 (b); R = *tert*-C_4H_9, R^1 = H (c); R = Ph, R^1 = H (d).

It was found with the help of results of chromatography-mass spectrometry analysis of products of interaction between equimolar amounts of acetone or acetophenone with C_2–C_5 mercaptans, formaldehyde and sodium hydroxide that in their composition there are 56-62% of disubstituted LXXXVIIIa (or LXXXVIIId) and 15-18% of monosubstituted derivatives LXXXVIIa (or LXXXVIId).

The synthesized γ-ketosulfides LXXVIIa-f and LXXXVIIIa-c represent high boiling liquids of light yellow, well soluble in the majority of organic solvents and insoluble in water. Structures of these compounds were confirmed by spectral methods. So, the distinguishing feature of IR spectra of mentioned products are stretching vibrations of group >C=O which are characterized by intensive absorption bands at 1715-1705cm^{-1} [167, 209].

The contents of individual components in reaction products mixture were determined for some of synthesized β-alkylthioketones and β,β'-di(alkylthio)ketones with the help of chromatography-mass spectrometry (Table 4.7).

Table 4.7. Individual compositions of mixtures of synthesized β-alkylthioketones and β,β'-di(alkylthio)ketones(in accordance with chromatography-mass spectrometry data)

Alk	Content of component in compounds mixture, %			
	LXXXVIIa	LXXXVIIc	LXXXVIIe	LXXXVIIIb
C_2H_5	3.5	3.3	3.2	3.1
n-C_3H_7	22.0	21.7	22.2	20.8
iso-C_3H_7	40.9	41.2	40.7	41.0
n-C_4H_9	5.4	5.5	5.6	6.0
iso-C_4H_9	22.0	22.3	22.3	23.7
n-C_5H_{11}	1.7	1.5	1.8	1.5
iso-C_5H_{11}	4.5	4.5	4.2	3.9

4.3.2. Transformations of N-, O-Containing Derivatives of β-Alkylthioketones

4.3.2.1. Synthesis and Transformations of N-Containing Derivatives of β-Alkylthioketones

We synthesized the elementary N-containing derivatives pf β-alkylthioketones: azomethines, hydrazones, N-acylhydrazones and thiosemicarbazones.

Azomethines discovered 140 years ago by Kh. Shiff which in many cases are obtained with the high yield from aldehydes, ketones and compounds containing primary amino-group found wide application in organic synthesis, industry and medicine [258-260]. However in spite of essential number of publications devoted to Kh, Shiff bases there are few data in literature on azomethines of ketosulfides number.

Firstly we studied the interaction of 4-alkylthiobutanones-2 LXXXVIIa, 5-alkylthio-2,2-dimethylpentanones-3 LXXXVIIc and 3-alkylthio-1-phenylpropanones-1 LXXXVIId with some primary amines of aliphatic, aromatic and heterocyclic types: *n*-hexylamine, aniline, *o*-aminophenol, *n*-chloroaniline, *cyclo*-hexylamine, 4-hydroxy-3,5-di-*tert*-butylaniline, 2-aminopyridine and furfurylamine.

Usually azomethines are obtained by interaction of equimolar amounts of carbonyl compound and primary amine in alcohol solution at moderate heating [258]. However it turned out that in the case of β-ketosulfides LXXXVIIa, c, d at mentioned conditions reaction proceeds normally only with *n*-hexyl-, cyclo-hexyl- and furfurylamines. Desired azomethines LXXXIXa, b, f, h, i, l, o were synthesized with the yield 63-90% by boiling of equimolar amounts of mentioned amines in ethanole (or propanol-2) for 4-5 hours (Table 4.8).

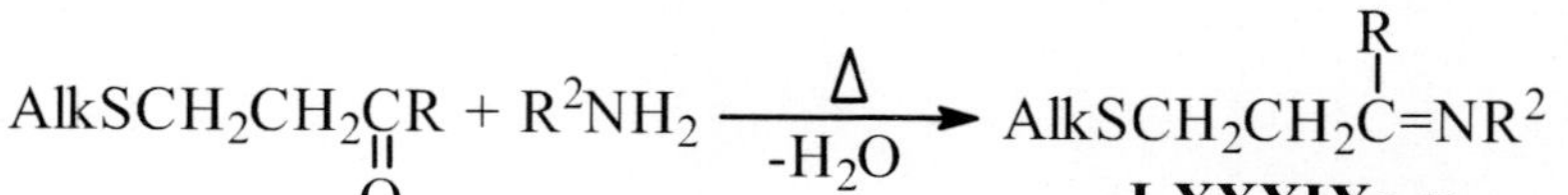

LXXXIXa-p

Alk = alkyl C_2-C_5; R = CH_3 (LXXXVIIa, LXXXIXa-f), *tert*-C_4H_9 (LXXXVIIc, LXXXIXh-l), Ph (LXXXVIId, LXXXIXm-p); R^2 = *n*-C_6H_{13} (LXXXIXa, h), *cyclo*-C_6H_{11} (LXXXIXb, i), Ph (LXXXIXc), 2-HOC_6H_4 (LXXXIXd, j, n), 4-ClC_6H_4 (LXXXIXm), 4-HO-3,5-(*tert*-C_4)$_2$ (LXXXIXe, k), furfuryl (LXXXIXf, l, o), pyridile-2 (LXXXIXg, p)

Table 4.8. Constants, yield and data of elementary analysis of β-alkylthioketones $AlkSCH_2CH_2CR = NR^2$

No.	R	R^2	Yield, %	$T_{melt.}$, °C (solvent for crystallization)	n_D^{20}	S, mass %
LXXXIXa	CH_3-	*n*-C_6H_{13}-	84	oil*)	1.5439	13.81
LXXXIXb	CH_3-	*cyclo*-C_6H_{11}	78	oil*)	1.5512	14.74
LXXXIXc	CH_3-	Ph-	88	74-79 (benzene-heptane, 3:1)	-	14.02
LXXXIXd	CH_3-	2-HOC_6H_4-	73	122-127 (dioxane-heptane, 1:2)	-	12.92
LXXXIXe	CH_3-	4-HO-3,5-(*tert*-$C_4H_9)_2$ C_6H_2-	70	88-94 (petroleum-ether)	-	8.85
LXXXIXf	CH_3-	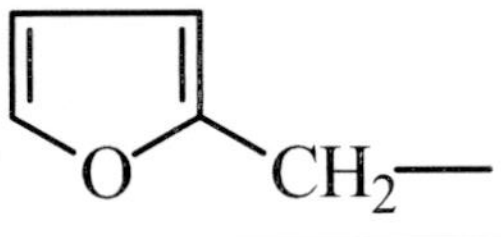	90	oil*)	1.5485	13.41
LXXXIXg	CH_3-	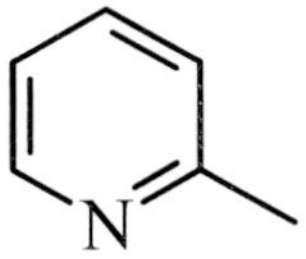	80	84-89 (ethylacetate)	-	13.78
LXXXIXh	*tert*-C_4H_9	*n*-C_6H_{13}	69	oil*)	1.5549	11.31
LXXXIXi	*tert*-C_4H_9	*cyclo*-C_6H_{11}	63	oil*)	-	12.05
LXXXIXj	*tert*-C_4H_9	2-HOC_6H_4-	56	106-111 (dioxane-heptane, 1:1)	-	11.82
LXXXIXk	*tert*-C_4H_9	4-HO-3,5-(*tert*-$C_4H_9)_2$ C_6H_2	58	70-77 (petroleum-ether)	-	8.27
LXXXIXl	*tert*-C_4H_9	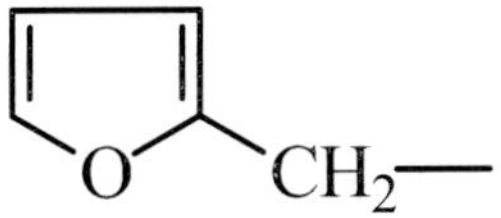	61	oil*)	1.5533	11.40
LXXXIXm	Ph-	4-ClC_6H_4-	71	87-94 (dioxane-heptane, 1:1)	-	9.65

Table 4.8. Continued

No.	R	R^2	Yield, %	$T_{melt.}$, °C (solvent for crystallization)	n_D^{20}	S, mass %
LXXXIXn	Ph-	2-HOC_6H_4-	57	135-141 (ethylacetate)	-	10.79
LXXXIXo	Ph-	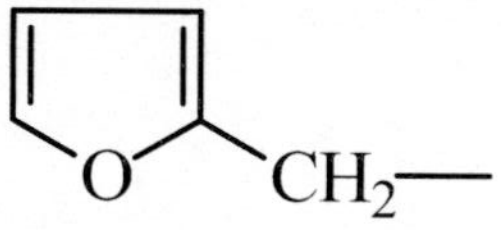	64	94-100 (petroleum-ether)	-	11.12
LXXXIXp	Ph-	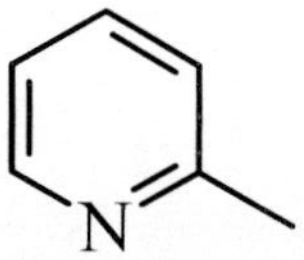	66	92-97 (dioxane-heptane, 1:2)	-	10.88

Note: *) The compounds were purified by column chromatography on silica gel.

Attempts of reception of corresponding azomethines from aromatic amines and 2-aminopyridine in the majority of cases were unsuccessful and led to low yield of azomethines. Nevertheless we showed by variation of conditions of experiments that azomethines LXXXIXc-e, j, k, m, n, p may be synthesized with the yield 56-88% by boiling of equimolar amounts of reagents in alcohol in the presence of catalytic amounts of acetic acid or *n*-toluenesulfoacid.

The noted differences in proceeding of reactions for aliphatic and aromatic amines are obviously conditioned by relatively low reactivity of carbonyl group in β-alkylthioketones. Basicity of initial amines also essentially influences on its ability to react with these ketones. For example amines with low basicity in particular *n*-nitroaniline (pK_a 1,02) and *o*-nitroaniline (pK_a -0,29) at conditions mentioned above do not interact with β-alkylthioketones.

We noticed that yield of azomethines depended also on nature of substituent R in initial ketone. So, on the base of ketones LXXXVIIc, d containing *tert*-C_4H_9 or phenyl groups at reactive site the corresponding azomethines are formed with lower yields (Table 4.8) than on the base of 4-alkylthiobutanones-2 LXXXVIIa (R = CH_3).

Under boiling (5-6 hours) of β-alkylthioketones LXXXVIIa-f with thio-semicarbazide and 4-phenylthiosemicarbazide in propanol-2 in the presence of catalytic amounts of acetic acid the crystal 4-R^2-thiosemicarbazones LXXXXa-j were synthesized with high yield (Table 4.9).

Table 4.9. Constants, yield and data of elementary analysis of thiosemicarbazones and 4-phenylthiosemicarbazones of β-alkylthioketones $AlkSCH_2CHR^1CR{=}NNHCSNHR^2$

No.	R	R^1	R^2	Yield, %	$T_{melt.}$, °C (solvent for crystallization)	S, mass %
LXXXXa	CH_3-	H-	H-	88	168-173 (propanol-2)	27.62
LXXXXb	CH_3-	H-	Ph-	82	142-148 (propanol-2)	20.65
LXXXXc	CH_3-	CH_3-	H-	78	152-156 (ethylacetate)	27.11
LXXXXd	CH_3-	CH_3-	Ph-	74	138-144 (dioxane-water. 1:1)	19.91
LXXXXe	*tert*-C_4H_9-	H-	H-	73	140-147 (ethanol)	23.72
LXXXXf	Ph-	H-	H-	81	170-175 (ethanol)	21.94
LXXXXg	Ph-	H-	Ph-	72	161-166 (propanol-2–water. 3:1)	18.10
LXXXXh	furyl-2-	H-	H-	86	138-144 (ethylacetate)	22.81
LXXXXi	furyl-2-	H-	Ph-	75	122-127 (ethanol-water. 2:1)	18.78
LXXXXj	-$(CH_2)_3$-		H-	70	150-156 (propanol-2)	24.98

The corresponding N-acylhydrazones of ketones LXXXXIa-q were obtained at analogous conditions from β-alkylthioketones LXXXVIIa, c, d and hydrazides of carbonic acids (Table 4.10).

Table 4.10. Constants, yield and data of elementary analysis of N-acylhydrazines of β-alkylthioketones $AlkSCH_2CH_2CR{=}NNHCOR^2$

No.	R	R^2	Yield, %	$T_{melt.}$, °C (solvent for crystallization)	S, mass %
LXXXXIa	CH_3-	Ph-	88	56-61 (petroleum-ether)	12.24
LXXXXIb	CH_3-	4-$NO_2C_6H_4$-	82	87-93 (propanol-2–water, 1:1)	10.82
LXXXXIc	CH_3-	furyl-2-	73	oil*⁾ (n_D^{20} 1.5722)	11.64
LXXXXId	CH_3-	4-HO-3,5-(*tert*-$C_4H_9)_2C_6H_2C_2H_4$-	74	oil*⁾ (n_D^{20} 1.5810)	8.04
LXXXXIe	CH_3-	4-HO-3,5-(*tert*-$C_4H_9)_2$-$C_6H_2SC_2H_4$-	68	oil*⁾ (n_D^{20} 1.5767)	14.22
LXXXXIf	CH_3-	5-nitrofuryl-2	70	109-115 (ethanol-water, 3:1)	9.94
LXXXXIg	CH_3-	pyridile-3	77	100-105 (benzene-heptane, 2:1)	11.24
LXXXXIh	*tert*-C_4H_9	4-$NO_2C_6H_4$-	65	92-97 (ethanol-water, 1:1)	9.41
LXXXXIi	*tert*-C_4H_9	4-HO-3,5-(*tert*-$C_4H_9)_2$ $C_6H_2C_2H_4$-	64	oil*⁾ (n_D^{20} 1.5788)	7.14
LXXXXIj	*tert*-C_4H_9	4-HO-3,5-(*tert*-$C_4H_9)_2C_6H_2SC_2H_4$-	68	oil*⁾ (n_D^{20} 1.5824)	13.06
LXXXXIk	*tert*-C_4H_9	furyl-2	80	oil*⁾ (n_D^{20} 1.5692)	11.08
LXXXXIl	*tert*-C_4H_9	5-nitrofuryl-2	71	118-124 (ethanol-water, 1:1)	8.92
LXXXXIm	Ph-	4-HO-3,5-(*tert*-$C_4H_9)_2C_6H_2C_2H_4$-	81	58-64 (petroleum-ether)	7.03
LXXXXIn	Ph-	4-HO-3,5-(*tert*-$C_4H_9)_2C_6H_2SC_2H_4$-	74	75-81 (petroleum-ether)	11.90
LXXXXIo	Ph-	furyl-2	63	65-71 (benzene-heptane, 1:1)	10.24
LXXXXIp	Ph-	5-nitrofuryl-2	70	135-143 (propanol-2)	8.45
LXXXXIq	Ph-	pyridile-3	78	115-121 (ethanol-water, 1:1)	9.67

Note: *⁾ The compounds were purified by column chromatography on silica gel.

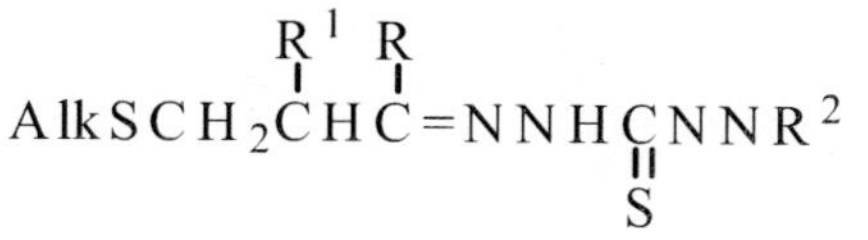

LXXXXa-j

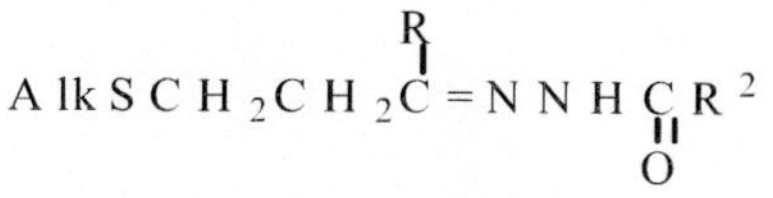

LXXXXIa-q

Alk = alkyl C_2-C_5;

90: $R = CH_3$, R^1 = H (a, b); $R = R^1 = CH_3$ (c, d); R = *tert*-C_4H_9, R^1 = H (e); R = Ph, R^1 = H (f, g); R = furyl-2, R^1 = H (h, i); $R + R^1 = (CH_2)_3$ (j); R^2 = H (a, c, e, f, h, j), Ph (b, d, g, i);

91: $R = CH_3$ (a-g), *tert*-C_4H_9 (h-l), Ph (m-q); R^2 = Ph (a), 4-$NO_2C_6H_4$ (b, h), 4-HO-3,5-(*tert*-$C_4H_9)_2C_6H_2CH_2CH_2$ (c, i, m), 4-HO-3,5-(*tert*-$C_4H_9)_2C_6H_2SCH_2CH_2$ (d, j, n), furyl-2 (e, k, o), 5-nitrofuryl-2 (f, l, p), pyridile-3 (g, q)

One of the most well studied chemical transformations of azomethines is their reduction into secondary amines [258]. Reduction of Kh. Shiff basis is considered as classical method of secondary amines synthesis in comparison with others more complex methods [261, 262]. For reduction of azomethine group they usually used complex hydrides of various metals or treatment by H_2 in the presence of Ni-, Pt- or Pd-catalysts [262-265].

With the aim of reception of secondary γ-alkylthioamines we studied reduction of derivatives of β-alkylthioketones LXXXIXa, b, e, f, i, k, l, o by sodium boronhydride selection of which was conditioned by its well known [263, 264] high efficiency and selectivity as reducing agent of C=N– bond in azomethines of various structures. We established in a number of experiments that reduction of azomethines should be carried out at double excess of sodium boronhydride by boiling in ethanol for 4-5 hours. At these conditions N-substituted 1-amino-3-alkylthio-1-R-propanes LXXXXIIa-h are formed with the yield 65-82% (Table 4.11).

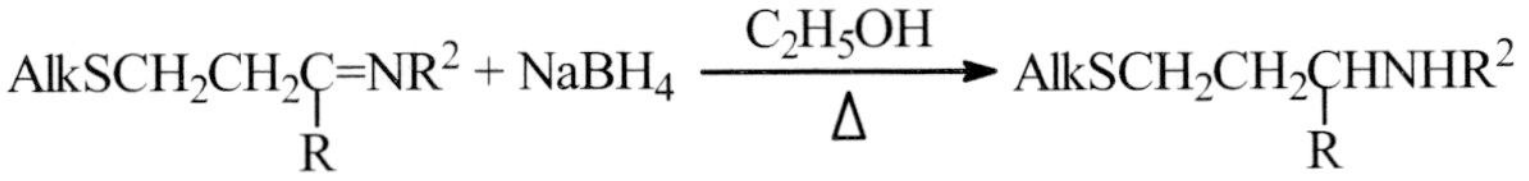

LXXXIXa, b, e, f, i, k, l, o **LXXXXIIa-h**

Alk = alkyl C_2-C_5;

R = CH_3 (LXXXVIIa, b, e, f; LXXXXIIa-d), *tert*-C_4H_9 (LXXXVIIi, k, l; LXXXXIIe-g), Ph (LXXXVIIo, LXXXXIIh);

R^2 = *n*-C_6H_{13} (LXXXVIIa, LXXXXIIa), *cyclo*-C_6H_{11} (LXXXVIIb, i; LXXXXIIb, e), 4-HO-3,5-di-*tert*-C_4-C_6H_3 (LXXXVIIe, k; LXXXXIIc, f), furfuryl (LXXXVIIf, l, o; LXXXXIId, g, h)

We found that under reduction in methanol medium (or propanol-2) the yield of desired secondary amines was decreased by 10-15% and increase of sodium boronhydride amount up to 3-4equivalents and duration of reduction up to 6-8 hours did not lead to noticeable increase of yield of γ-aminosulfides LXXXXIIa-h.

Synthesis of elementary representatives of γ-hydroxysulfides (3-alkylthiopropanols-1 Ib) by alkylation of C_2–C_5 mixture by 3-chloropropanols is described in Section 2.2.1. Such secondary alcohols, γ-hydroxysulfides LXXXXIIIa-e were synthesized in presented section of work by reduction of corresponding β-alkylthioketones LXXXVIIa-f with high yield (Table 4.12). Reaction of reduction was realized under the action of sodium boronhydride in aqueous ethanol at 40-45°C for 2-3 hours. At analogous conditions β,β'-di(alkylthio)ketones LXXXVIIIa-d were transformed into γ-hydroxy-*bis*-sulfides LXXXXIVa-d with the yield 79-85%.

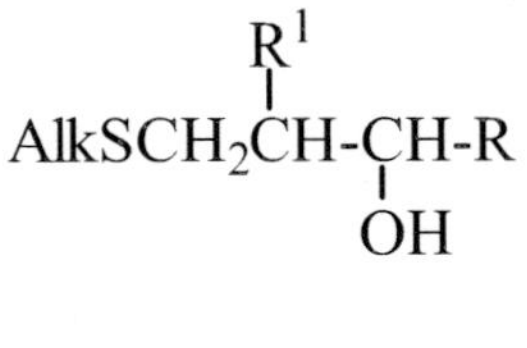

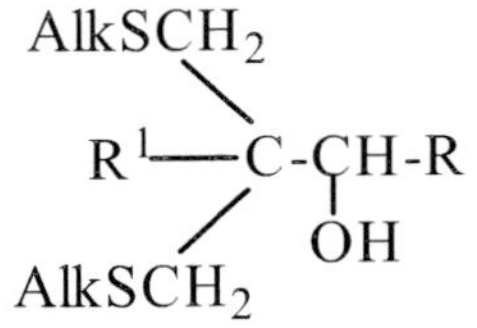

LXXXXIIIa-f **LXXXXIVa-d**

Alk = alkyl C_2-C_5;

R = CH_3, R^1 = H (LXXXXIIIa, LXXXXIVa); R = R^1 = CH_3 (LXXXXIIIb, LXXXXIVb); R = *tert*-C_4H_9, R^1 = H (LXXXXIIIc, LXXXXIVc); R = Ph, R^1 = H (LXXXXIIId, LXXXXIVd); R = furyl-2, R^1 = H (LXXXXIIIe); R + R^1 = $(CH_2)_3$ (LXXXXIIIf)

Physical-chemical characteristics of nitrogen-containing derivatives of β-alkylthioketones LXXXIX-LXXXXI, γ-aminosulfides LXXXXIIa-h, γ-hydrosulfides LXXXXIIIa-e and γ-hydroxybissulfides LXXXXIVa-d are presented in Tables 4.8-4.12. structures of synthesized compounds were confirmed by the data of IR and PMR spectroscopies and also elementary analysis.

Table 4.11. Constants, yield and data of elementary analysis of N-substituted 1-amino-1-R-3-alkylthiopropanes $AlkSCH_2CH_2CHRNHR^2$

No.	R	R^2	Yield, %	$T_{boil.}$, °C (millimeters of mercury)	n_D^{20}	d_4^{20}, g/cm^3	S, mass %
LXXXXIIa	CH_3-	*n*-C_6H_{13}-	80	64-88 (2)	1.5133	0.9415	12.97
LXXXXIIb	CH_3-	*cyclo*-C_6H_{11}-	72	79-102 (2)	1.5184	0.9610	13.72
LXXXXIIc	CH_3-	4-HO-3,5-(*tert*-C_4H_9)$_2$-C_6H_2-	64	oil*)	1.5304	0.9485	9.12
LXXXXIId	CH_3-	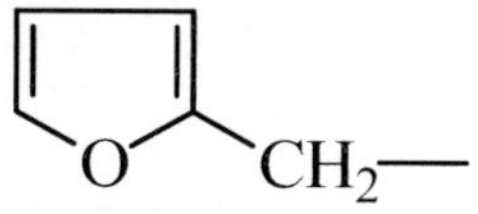	75	129-147 (2)	1.5281	0.9554	13.65
LXXXXIIe	*tert*-C_4H_9	*cyclo*-C_6H_{11}-	62	121-140(1)	1.5224	0.9459	11.84
LXXXXIIf	*tert*-C_4H_9	4-HO-3,5-(*tert*-C_4H_9)$_2$-C_6H_2-	65	oil*)	1.5381	1.0118	7.62
LXXXXIIg	*tert*-C_4H_9	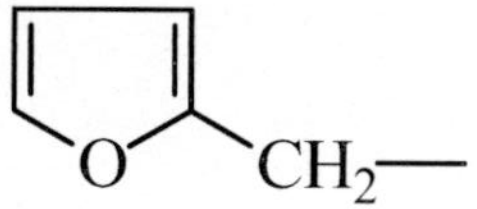	78	149-165 (1)	1.5342	0.9785	11.12
LXXXXIIh	Ph-	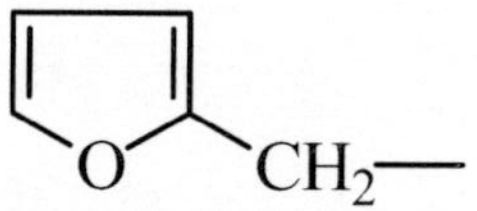	67	165-180 (2)	1.5432	1.0237	11.02

Note: *) The compounds were purified by column chromatography on silica gel.

Table 4.12. Constants, yield and data of elementary analysis of γ-hydroxysulfides $AlkSCH_2CHR^1CH(OH)R$ and γ-hydroxy-*bis*-sulfides $(AlkSCH_2)_2CR^1CH(OH)R$

No.	R	R^1	Yield, %	$T_{boil.}$, °C (millimeters of mercury) or $T_{melt.}$, °C (solvent for crystallization)	n_D^{20}	d_4^{20}, g/cm³	S, mass %
LXXXXIIIa	CH_3-	H-	80	80-103 (2)	1.4764	0.9384	20.87
LXXXXIIIb	CH_3-	CH_3-	77	104-125 (2)	1.4802	0.9335	18.68
LXXXXIc	*tert*-C_4H_9	H-	78	126-145 (2)	1.4862	0.9416	16.75
LXXXXIIId	Ph-	H-	84	43-48 (ethanol-water, 1:1)	-	-	15.10
LXXXXIIIe	фурил-2-	H-	82	138-151 (1)	1.5135	0.9556	14.62
LXXXXIIIf	-$(CH_2)_3$-		77	115-135 (1)	1,4926	0.9330	18.09
LXXXXIVa	CH_3-	H-	82	120-131 (1)	1.5230	0.9416	26.04
LXXXXIVb	CH_3-	CH_3-	85	128-147 (1)	1.5266	0.9435	23.99
LXXXXIVc	*tert*-C_4H_9	H-	81	39-45 (heptane-MEK, 5:1)	-	-	22.29
LXXXXIVd	Ph-	H-	79	57-63 (ethanol-water, 2:1)	-	-	20.87

Table 4.13. Constants, yield and data of elementary analysis of derivatives of 1,3-dioxolane, 1,3-oxothiolane and 1,3-dithiolane containing alkylsulfide groups

LXXXXIIIa-g
LXXXXIVa-g
LXXXXVa-d

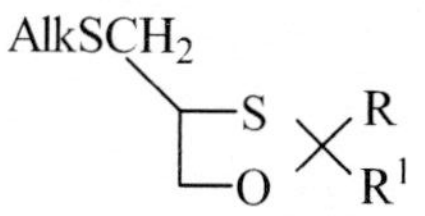

LXXXXVIa-d

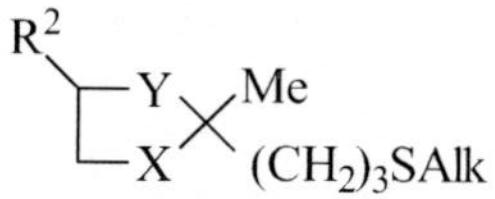

LXXXXIVa, b
Ca, b, CI

Compound	X	Y	R	R^1	R^2	Yield, %	$T_{boil.}$, °C (millimeters of mercury) or $T_{melt.}$, °C (solvent for crystallization)	n_D^{20}	d_4^{20}, g/cm^3	S, mass %
LXXXXVa		O	CH_3-	H-	H-	88	65-88 (1)	1.4568	1.0862	16.79
LXXXXVb	O	O	CH_3-	CH_3-	H-	83	70-93 (1)	1.4554	1.0784	15.76
LXXXXVc	O	O	*tert*-C_4H_9	H-	H-	56	103-123 (2)	1.4592	1.0890	13.81
LXXXXVd	O	O	Ph-	H-	H-	58	119-138 (2)	1.4608	1.884	12.80
LXXXXVe	O	O	CH_3-	H-	$HOCH_2$-	82	124-143 (1)	1.4534	1.0760	14.42
LXXXXVf	O	O	CH_3-	CH_3-	$HOCH_2$-	76	137-154 (1)	1.4581	1.0910	13.84
LXXXXVg	O	O	Ph-	H-	$HOCH_2$-	64	58-63 (ethanol-water, 1:1)	-	-	11.37
LXXXXVIa	S	O	CH_3-	H-	H-	93	75-98 (1)	1.4612	0.9912	30.81
LXXXXVIb	S	O	CH_3-	CH_3-	H-	74	179-111 (1)	1.4590	1.0044	28.94
LXXXXVIc	S	O	*tert*-C_4H_9	CH_3-	H-	58	105-125 (1)	1.4624	1.0946	24.77
LXXXXVId	S	O	Ph-	H-	H-	66	133-151 (1)	1.4642	1.1010	23.93
LXXXXVIe	S	O	CH_3-	H-	Ph-	92	43-48 (petroleum-ether)	-	-	23.34
LXXXXVIf	S	O	*tert*-C_4H_9	H-	Ph-	71	38-44 (гексан)	-	-	20.04
LXXXXVIg	S	O	Ph-	H-	Ph-	60	67-72 (ethanol-water, 1:1)	-	-	18.58
LXXXXVIIa	S	S	CH_3-	H-	H-	87	85-107 (2)	1.4652	1.1056	43.08
LXXXXVIIb	S	S	CH_3-	CH_3-	H-	73	99-119 (2)	1.4668	1.1072	41.51
LXXXXVIIc	S	S	*tert*-C_4H_9-	H-	H-	57	111-130 (1)	1.4677	1.1088	36.57
LXXXXVIId	S	S	Ph-	H-	H-	53	126-144 (1)	1.4692	1.1118	33.44
LXXXXVIIIa	-	-	H-	H-	-	88	48-74 (3)	1.4338	0.9814	20.40
LXXXXVIIIb	-	-	Ph-	H-	-	84	54-79 (2)	1.4404	1.0081	13.85

Table 4.13. (Continued)

Compound	X	Y	R	R^1	R^2	Yield, %	$T_{boil.}$, °C (millimeters of mercury) or $T_{melt.}$, °C (solvent for crystallization)	n_D^{20}	d_4^{20}, g/cm^3	S, mass %
LXXXXVIIIc	-	-	CH_3-	CH_3-	-	80	51-76 (2)	1.4389	0.9902	16.61
LXXXXVIIId	-	-	CH_3-	C_2H_5-	-	82	66-80 (2)	1.4394	0.9918	15.86
CIa	O	O	-	-	H	76	88-110 (2)	1.4537	1.0108	16.34
CIb	O	O	-	-	$HOCH_2$	70	135-152 (1)	1.4582	1.0134	13.66
CIIa	S	O	-	-	H	75	107-126 (1)	1.4610	1.0122	28.81
CIIb	S	O	-	-	Ph	72	51-55 (ethanol-water, 1:1)	-	-	21.49
CIII	S	S	-	-	H	71	116-134 (1)	1.4644	1.0158	40.90

In the IR spectra of compounds LXXXIXa-p there are intensive absorption bands in the region 1625-1605cm^{-1} characteristic for stretching vibrations of C=N group in azomethines [22, 23, 161]. At the same time in spectra of thiosemicarbazones LXXXXa-j and N-acylhydrazones LXXXXIa-q the frequencies of absorption of given group are shifted to high frequency region and are revealed at 1660-1635cm^{-1} [222, 223]. In spectra of N-acylhydrazones LXXXXIa-q the intensive absorption maximums are also revealed at 1690-1655cm^{-1} relating to stretching vibrations of carbonyl group (the band of "amide I" type) [102, 105]. In spectra of thiosemicarbazones LXXXXa-j there is a number of bands of variable intensity at 1555-1540, 1460-1445 and 1330-1320 cm^{-1} characteristic for stretching vibrations of C=S group in thioamide fragment [167, 211].

In high frequency region of spectra of γ-aminosulfides LXXXXa-h weak absorption band was fixed in the region 3350-3310cm^{-1} typical for stretching vibrations of N–H in secondary amines [167, 209, 211]. Deformation vibrations of these bonds are characterized by absorption bands of variable intensity in the interval 1545-1515cm^{-1},and stretching vibrations of C–N are presented by a number of strong bands at 1340-1285cm^{-1} characteristic for secondary amines [167, 211].

In high frequency region of spectra of secondary alcohols LXXXXIIIa-f and LXXXXIVa-d there are broad absorption bands at 3465-3330cm^{-1} peculiar to stretching vibrations of associated HO-groups in aliphatic alcohols [167, 209]. In spectra of these compounds the absorption maximums in the regions 1420-1385, 1335-1320 (planar deformation vibrations of O–H) and 1125-1060cm^{-1} ($\nu_{C\text{-}O}$) were also observed [209].

In PMR spectra of thiosemicarbazones LXXXXa-j and N-acylhydrazones LXXXXIa-q signals of protons of NH-group are revealed as broadened singlets in weak field (8.90-10.85m.p.). The broadened signals in the interval 6.18-6.67m.p. characterize protons of secondary amino-groups in spectra of γ-aminosulfides LXXXXIIa-h. In PMR spectra of alcohols LXXXXIIIa-f and LXXXXIVa-d signals of hydroxyl proton are presented by broadened singlets at 5.50-5.84m.p..

4.3.2.2. Synthesis of derivatives of 1,3-dioxolane, 1,3-oxathiolane and 1,3-dithiolane containing alkylsulfide groups

At present 1,3-diheterocyclopentanes (1,3-dioxolanes, 1,3-oxathiolanes, 1,3-dithiolanes, oxazolidines, imidazolidines, thiazolidines) present great interest in connection with prospects of their application as plasticizers and light-stabilizers of polymers, extracting agents, herbicides, regulators of plants growth, fungicides, veterinary and medical preparations [58,173,212,266-268]. Functional substituted derivatives of 1,3-diheterocyclopentanes including those with sulfur-containing groups may also present essential importance. In particular, it is known that 4-alyklthiomethyl- [269, 270] and 2-(β-alkylthioethyl)-1,3-dioxolanes possess high antibacterial activity.

In development of these investigations in presented work we synthesized derivatives of 1,3-dioxolane, 1,3-oxathiolane and 1,3-dithiolane containing alkylsulfide groups at α-, β- and γ-positions of side chain. Mixture of C_2–C_5 mercaptans and β-alkylthioketones LXXXVIIa-d described above were used as key compounds. So, in accordance with method reported in [212] by condensation of equimolar amounts of mentioned ketones with ethylene glycol, glycerin, 2-mercaptoethanol, 2-mercapto-1-phenylethanol or 1,2-ethanedithiole in the presence of cationite KU-2x8 (10-15 mass %) in benzene (or toluene) with azeotrope removal of formed water 2-R-2-(2-alkylthio-1-R[1]-ethyl)-4-R[2]-1,3-diozolanes LXXXXVa-g, 1,3-oxathiolanes LXXXXVIa-g and 1,3-dithiolanes LXXXXVIIa-d were synthesized.

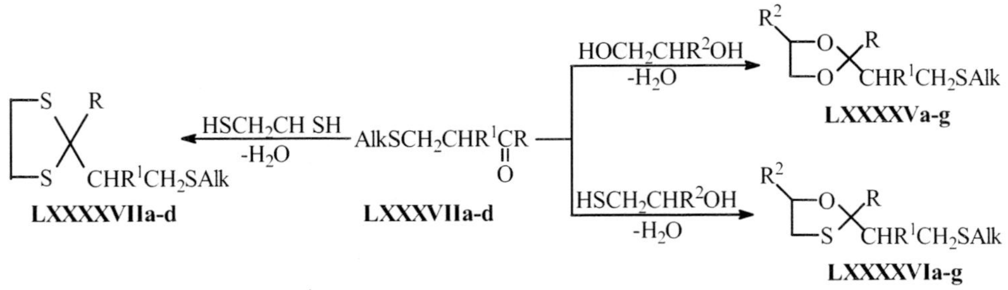

Alk = alkylл C_2-C_5; R = CH_3, R^1 = H (LXXXVIIa, LXXXXVa, e, LXXXXVIa, e, LXXXXVIIa); R = R^1 = CH_3 (LXXXVIIb, LXXXXVb, f, LXXXXVIb, LXXXXVIIb); R = *tert*-C_4H_9, R^1 = H (LXXXVIIc, LXXXXVc, g, LXXXXVIc, f, LXXXXVIIc); R = Ph, R^1 = H (LXXXVIId, LXXXXVd, LXXXXVId, g, LXXXXVIId); R^2 = H (LXXXXVa-d, LXXXXVIa-d), $HOCH_2$-(LXXXXVe-g), Ph (LXXXXVIe-g).

When varying conditions in a number of experiments we established that reactions duration and yield of desired products depended mainly on the structure of initial carbonyl compounds. For example application of 4-alkylthio-butanones-2 LXXXVIIa in synthesis gives heterocycles LXXXXVa, e, LXXXXVIa, e, LXXXXVIIa with the yield 82-93% (Table 4.13) after boiling in benzene for 3-4 hours, and initial 4-alkylthio-3-methylbutanones-2 LXXXVIIb is necessary to boil for 7-8 hours to receive 1,3-dioxalanes LXXXXVb, f, 1,3-oxathiolanes LXXXXVIb and 1,3-ditholanes LXXXXVIIb with the yield only 53-71%. Obviously the presence of volume *tert*-butyl and phenyl groups causes essential steric hindrances in cyclocondensation process.

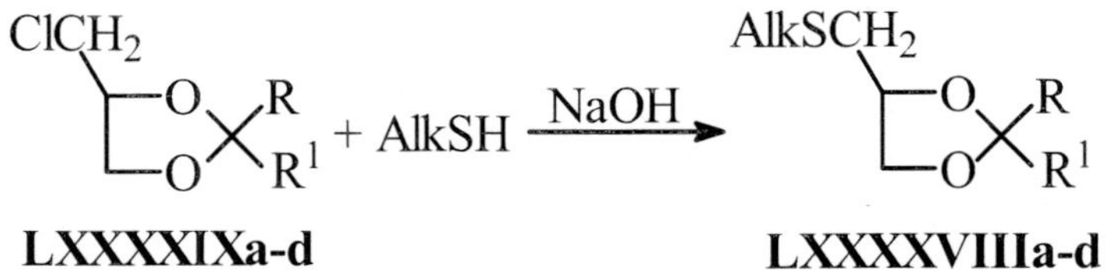

Alk = alkyl C_2-C_5; R = R^1 = H (a); R = Pr, R^1 = H (b); R = R^1 = CH_3 (c); R = CH_3, R^1 = C_2H_5 (d).

1,3-Dioxalanes containing alkylsulfide groups in α-position to cycle, (2-R-2-R[1]-4-alkylthiomethyl-1,3-dioxolanes LXXXXVIIIa-d) were synthesized by interaction of 2-R-2-R[1]-4-chloromethyl-1,3-dioxolanes LXXXXIXa-d with mixture of C_2–C_5 mercaptans and sodium hydroxide at conditions of inter-phase catalysis. When varying conditions of experiments we found that it was advisable to carry out reactions at molar ratio mercaptans : chloromethyl-1,3-dioxolanes : NaOH : catalyst (triphenylethylphosphoniumbromide or tetrabutylammoniumbromide) equal to 1 : 1.2 : 2 : 0.06 in solution of DMFA (or DMSO) at 60-70°C for 3-4 hours.

For reception of 1.3-diheterocyclopentanes with alkylsulfide groups in γ-position of side chain we used condensation of mixture of 5-alkylthiopentanones-2 C with corresponding 1,2-bifunctional reagents. Initial γ-alkylthioketones were synthesized with the yield 74% by alkylation of mercaptans fraction by 5-chloropentanone-2 in the presence of equivalent amount of alkali. 2-Methyl-2-(3-alkylthiopropyl)-4-R^2-1,3-dioxolanes CIa, b, 1,3-oxathiolanes CIIa, b and 1,3-dithiolanes CIII are formed with high yield (Table 4.13) under boiling of equimolar amounts of reagents for 5-6 hours in benzene in the presence of 10 mass % of cationite KU-2x8.

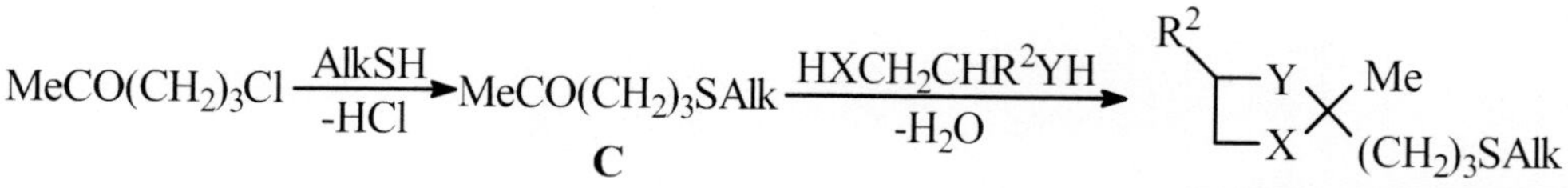

CIa, b; CIIa, b; CIII

Alk = alkyl C_2-C_5; X = Y = O (CIa, b); X = S, Y = O (CIIa, b); X = Y = S (CIII); R^2 = H (CIa, CIIa, CIII), $HOCH_2$- (CIb), Ph (CIIb

Structures of synthesized heterocyclic compounds were confirmed by data of IR and PMR spectroscopies. In IR spectra there are absorption maximums of variable intensity relating to vibrations of 1,3-diheterocyclo-pentane fragments [183, 213]: 1470-1450, 1445-1420 ("cutting" vibrations of CH_2), 1290-1275, 1085-1070 (ν_{cycle}), 1180-1165 and 1010-995cm^{-1}. In spectra of compounds LXXXXVa-g, LXXXXVIIIa-d and CIa, b there are also intensive absorption bands in the region 1075-1060cm^{-1} characteristic for vibrations of group –O–C–O– in 1,3-dioxolanes [57, 213]. In spectra of all synthesized compounds absorption of variable intensity in the region of wave length 2700-2600, 1430-1420 and 800-760 cm^{-1} relating to vibrations of C–S bond was fixed [167].

In PMR spectra of synthesized compounds in the region of resonance of protons of alkylthioalkyl fragments the signals with various values of chemical shift are observed. Signals of protons of methyl groups represent doublets or triplets at 1.04-1.16m.p.. Complex multiplets in the interval 1.22-1.64m.p. should be assigned to protons of methylene units, in the interval 1.70-1.88m.p. – to protons of methane groups, and signals shifted to the weak field (3.52-3.84m.p.) – to protons of CH_2– and CH–groups bonded with sulfur atom.

In spectra of 1,3-dioxolanes LXXXXVa-d and CIa, 1,3-oxathiolanes LXXXXVIa-d and CIIa and also 1,3-dithiolanes LXXXXVIIa-d and CIII signals of protons of methylene groups of cycle (4-CH_2 and 5-CH_2) are presented as expanded multiplets in the region 3.90-4.56m.p. [183, 271]. Signals of proton 4-H of cycle in spectra of compounds LXXXXVe-g, LXXXXVIe-g, LXXXXVIIIa-d, CIb and CIIb are observed in the shape of non-symmetrical multiplets in the interval 4.25-4.65m.p.. In spectra of 2-methyl-1,3-dioxolanes LXXXXVa, b, e, f, LXXXXVIIIb, d and CIa, b, 2-methyl-1,3-oxathiolanes LXXXXVIa, b, e and CIIa, b and also 2-methyl-1,3-dithiolanes LXXXXVIIa, b and CIII signals of protons of methyl groups are revealed as broadened singlets at 1.50-1.66m.p. that is characteristic for methyl groups in cycle systems [167, 183, 271].

The contents of individual components in mixtures of reaction products were determined for some of synthesized oxo- and oxathiolanes with the help of chromatography-mass spectrometry (Table 4.14).

Table 4.14. Individual composition of mixtures of synthesized heterocyclic compounds (in accordance with chromatography-mass spectrometry data)

Alk	Contents of components in compounds mixtures, %					
	LXXXXVa	LXXXXVf	LXXXXVIb	LXXXXVIIc	LXXXXVIIIa	CIII
C_2H_5	3.2	3.5	2.9	3.3	3.2	3.5
n-C_3H_7	21.7	20.9	22.0	21.6	20.5	21.3
iso-C_3H_7	40.5	41.0	39.6	38.5	40.9	39.8
n-C_4H_9	5.5	5.8	5.6	5.7	5.5	5.4
iso-C_4H_9	23.7	23.2	23.9	24.5	24.3	23.9
n-C_5H_{11}	1.5	1.6	1.7	1.8	1.6	1.7
iso-C_5H_{11}	3.9	4.0	4.3	4.6	4.0	4.4

4.3.2.3. Synthesis of Derivatives of Oxazolidine and Imidazolidine on the Base of β- and γ-Alkylthioketones

In development of investigations of condensation reactions of N-substituted 1-alkylthio-3-aminopropanols-2 LIb, c, e-h with various aldehydes described in Section 3.1.3 in given Section we synthesized oxazolidines containing alkylsulfide fragments in β- and γ-positions in relation to cycle with the use of β- and γ-alkylthioketones as key compounds. So, by condensation of equimolar amounts of N-substituted ethanolamines with 4-alkylthiobutanones-2 LXXXVIIa, 5-alkylthio-2,2-dimethylpentanones-3 LXXXVIIc, 3-alkylthio-1-phenylpropanones-1 LXXXVIId or with 5-alkylthiopentanones-2 C in benzene (or toluene) with azeotrope removal of water we obtained 2-R-2-(2-alkylthioethyl)- and CIVa-i and 2-methyl-2-(3-alkylthiopropyl)-3-R^1-oxazolidines CIVj-m.

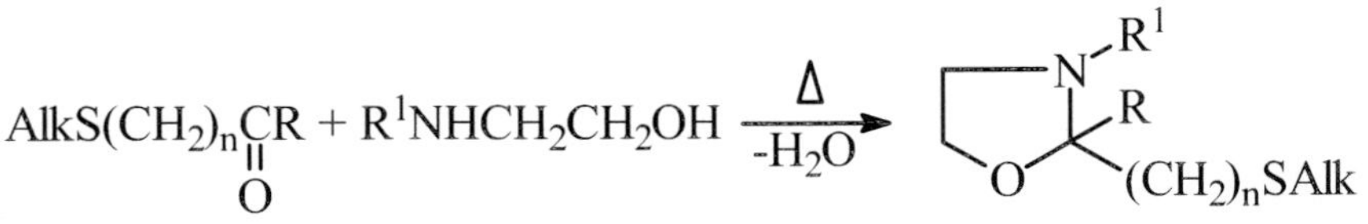

LXXXVIIa, c, d, C **CVIa-i**

Alk = alkyl C_2-C_5; n = 2 (LXXXVIIa, c, d; CIVa-i), 3 (C, CIVj-m); R = CH_3 (LXXXVIIa, C, CIVa-d, j-m), *tert*-C_4H_9 (LXXXVIIc, CIVe, f), Ph (LXXXVIId, CIVg-i); R^1 = *n*-C_4H_9 (CIVa, e, g, j), $PhCH_2$ (CIVb, h, k), 4-HO-3,5-(*tert*-C_4H_9)$_2$$C_6H_2CH_2$ (CIVc, l), furfuryl (CIVd, f, i, m)

When varied conditions of experiments we showed that duration of this reaction and yield of desired products were determined mainly by structure of initial ketones. For example on the base of β-alkylthioketones LXXVIIa and 5-alkylthiopentanones-2 C the corresponding oxazolidines CIVa-d, j-m are formed with the yield 80-90% after boiling in benzene for 7-8 hours. More prolonged boiling in toluene (12-16h) is necessary for synthesis of 2-*tert*-butyl-CIVe, f and 2-phenyloxazolidines CIIg-i which yield in this case does not exceed 54-63% (Table 4.15). Obviously as well as in the case of synthesis of dioxolanes described above the volume *tert*-butyl and phenyl group create steric hindrances in cyclization process.

Table 4.15. Constants, yield and data of elementary analysis of 2-R-2-(2-alkylthioethyl)-3-R^1- and 2-methyl-2-(3-alkylthiopropyl)-3-R^1-oxazolidine

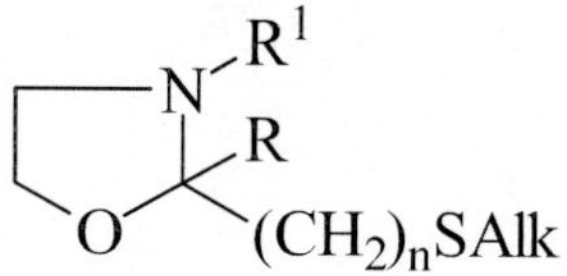

Compound	n	R	R^1	Yield, %	$T_{boil.}$, °C (millimeters of mercury) or $T_{melt.}$, °C (solvent for crystallization)	n_D^{20}	d_4^{20}, g/cm^3	S, mass %
CIVa	2	CH_3-	*n*-C_4H_9	82	80-101 (1)	1.5094	1.0732	12.14
CIVb	2	CH_3-	$PhCH_2$-	90	35-42 (hexane)	-	-	11.65
CIVc	2	CH_3-	4-HO-3,5-(*tert*-C_4H_9)$_2C_6H_2CH_2$-	85	48-56 (benzene)	-	-	7.40
CIVd	2	CH_3-		80	122-140 (2)	1.5235	1.1008	11.62
CIVe	2	*t*-C_4H_9	*n*-C_4H_9	54	69-92 (1)	1.5092	1.0748	11.07
CIVf	2	*t*-C_4H_9	Furyl-2-methyl-	56	133-151 (1)	1.5357	1.0838	9.98
CIVg	2	Ph-	*n*-C_4H_9	63	64-70 (ethanol-water, 2:1)	-	-	9.83
CIVh	2	Ph-	$PhCH_2$-	60	84-91 (ethanol-water, 2:1)	-	-	9.02
CIVi	2	Ph-	Furyl-2-methyl-	55	70-77 (ethanol-water, 1:1)	-	-	10.17
CIVj	3	CH_3-	*n*-C_4H_9	81	98-118 (2)	1.5233	1.0681	12.05
CIVk	3	CH_3-	$PhCH_2$-	85	52-57 (isooctane)	-	-	10.14
CIVl	3	CH_3-	4-HO-3,5-(*t*-C_4H_9)$_2C_6H_2CH_2$-	82	64-70 (petroleum-ether)	-	-	7.42
CIVm	3	CH_3-	Furyl-2-methyl-	80	140-157 (1)	1.5309	1.1014	10.77

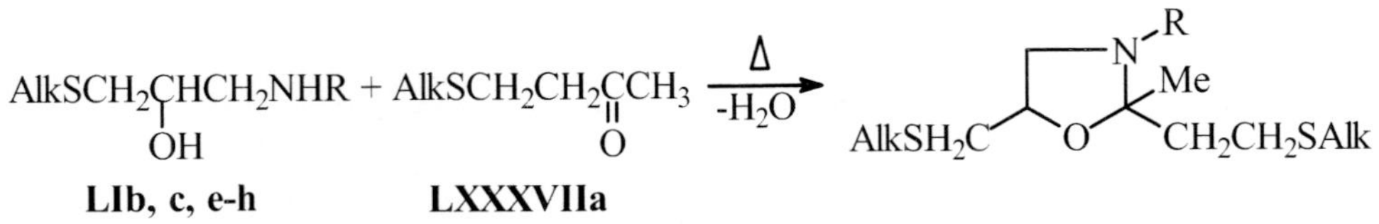

Alk = alkyl C_2-C_5; R = *n*-C_4H_9 (LIb, CVa), *n*-C_8H_{17} (LIc, CVb), $PhCH_2$ (LIe, CVc), $(CH_3)_2N(CH_2)_3$ (LIf, CVd), Ph (LIg, CVe), 4-HO-3,5-(*tert*-$C_4H_9)_2C_6H_2$ (LIh, CVf)

For reception of oxazolidines containing two alkylsulfide fragments we carried out condensation of N-substituted 1-alkylthio-3-amino-propanols-2 LIb, c, e-h described above with 4-alkylthiobutanones-2 LXXXVIIa. We used equimolar ratio of reagents and boiled them for 6-7 hours in water-less toluene with azeotrope removal of formed water and obtained 2-methyl-2-(2-alkylthio)-3-R-5-alkylthiomethyloxazolidines CVa-f with the yield 63-74% (Table 4.16).

Imidazolidines except their application for isolation and identification of carbonyl compounds [173, 272] are of great interest from the point of view of their biological activity. At the same time these heterocycles are slightly studied until now and there are no data on imidazolidines with alkylsulfide fragments in literature at all. These facts conditioned the interest to synthesis of imidazolidines of such type in our work.

The most widely used method of reception of imidazolidines is condensation of N,N'-disubstituted ethylenediamines with carbonyl compounds [173, 272, 273] and it turned out to be convenient for synthesis of tetrasubstituted imidazolidines on the base of β- and γ-alkylthioketones. We established that under boiling (5-8 hours) of N,N'-disubstituted ethylenediamines with 4-alkylthiobutanones-2 LXXXVIIa, 5-alkylthio-2,2-dimethylpentanones-3 LXXXVIIc or 5-alkylthiopentanones-2 C (1 : 1, moles) in waterless benzene (10 mass % of Cationite KU-2x8) with azeotrope removal of formed water the 1,3-disubstituted 2-R-2-(2-alkylthioethyl)- CVIa-e and 2-methyl-2-(3-alkylthiopropyl)imidazolidines CVIf-h with the yield 50-68% are formed (Table 4.17).

Table 4.16. Constants, yield and data of elementary analysis of 2-methyl-2-(2-alkylthioethyl)-alkylthiomethyloxazolidines

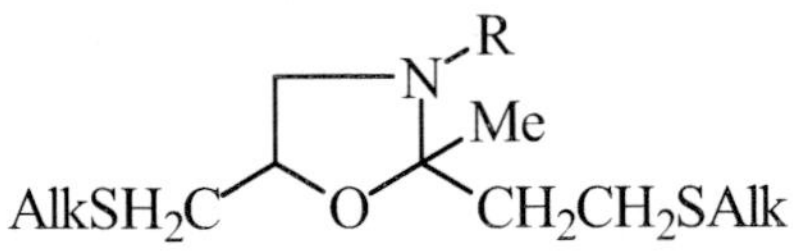

No.	R	Yield, %	$T_{boil.}$, °C (millimeters of mercury) or $T_{melt.}$, °C (solvent for crystallization)	n_D^{20}	d_4^{20}, g/cm^3	S, mass %
CVa	*n*-C_4H_9	63	132-150 (2)	1.5188	1.1109	18.87
CVb	*n*-C_8H_{17}-	68	48-58 (petroleum-ether)	-	-	16.16
CVc	$PhCH_2$-	74	82-90 (ethanol-water, 2:1)	-	-	17.31
CVd	$(CH_3)_2N(CH_2)_3$-	65	123-154 (1)	1.5354	1.1244	17.37
CVe	Ph-	71	94-103 (ethanol-water, 2:1)	-	-	18.42
CVf	4-HO-3,5-(*tert*-$C_4H_9)_2C_6H_2$-	70	72-80 (dioxane-water, 1:2)	-	-	13.45

Table 4.17. Constants, yield and data of elementary analysis of 2-R-2-(2-alkylthioethyl)- and 2-methyl-2-(3-alkylthiopropyl)-1,3-R_2^1-imidazolidines

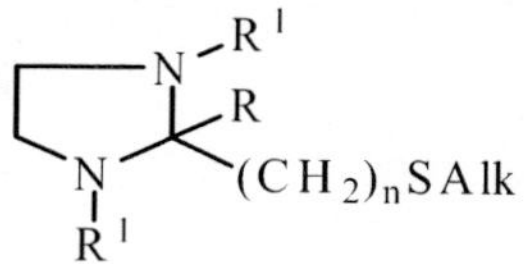

No.	n	R	R^1	Yield, %	$T_{melt.}$, °C (solvent for crystallization)	S, mass %
CVIa	2	CH_3-	CH_3-	62	oil*) (n_D^{20} 1.5732)	14.62
CVIb	2	CH_3-	$PhCH_2$-	68	52-60 (petroleum-ether)	8.34
CVIc	2	CH_3-	4-HO-(*tert*-C_4H_9)$_2$-C_6H_2-	64	42-50 (petroleum-ether)	5.09
CVId	2	*tert*-C_4H_9	Me-	54	oil*) (n_D^{20} 1.5764)	12.14
CVIe	2	*tert*-C_4H_9	$PhCH_2$-	50	oil*) (n_D^{20} 1.5814)	7.71
CVIf	3	CH_3-	CH_3-	60	масло*) (n_D^{20} 1.5748)	14.05
CVIg	3	CH_3-	$PhCH_2$-	58	39-44 (hexane)	7.95
CVIh	3	CH_3-	4-HO-(*tert*-C_4H_9)$_2$-C_6H_2-	55	63-70 (benzene-hexane, 1:1)	5.28

Note: *) The compounds were purified by column chromatography on silica gel.

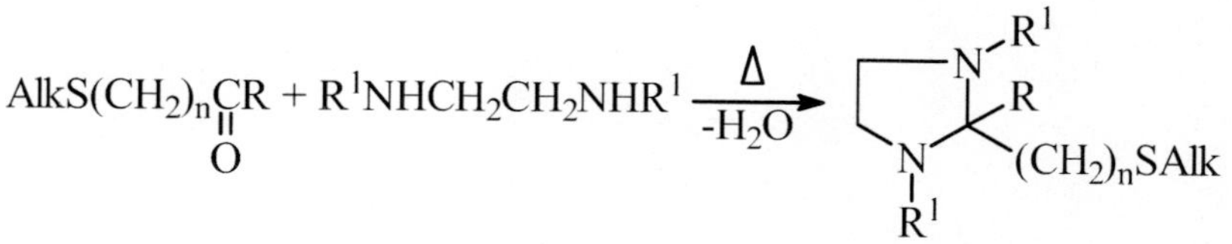

Alk = alkyl C_2-C_5; n = 2 (LXXXVIIa, c, CVIa-e), 3 (C, CVIf-h); R = CH_3 (LXXXVIIa, C, CVIa-c, f-h), *tert*-C_4H_9 (LXXXVIIc, CVId, e); R^1 = CH_3 (CVIa, d, f), $PhCH_2$ (CVIb, e, g), 4-HO-3,5-(*tert*-C_4H_9)$_2C_6H_2$ (XVIc, h)

We found that in the absence of acid catalyst imidazolidines CVIa-h are not formed in contrast to oxazolidines CIVa-i and CVa-f. For example after prolonged boiling (16-18 hours) of γ-alkylthioketones C with N,N'-dimethylethylendiamine in benzene only initial compounds were observed in reaction mixture (in accordance with gas-liquid chromatography).

Physical-chemical characteristics of synthesized derivatives of oxazolidine and imidazoline are presented in Tables 4.15-4.17. structures of these compounds were confirmed by the data of IR and PMR spectroscopy and also by the data of elementary analysis. In IR spectra of heterocyclic compounds CIVa-i and CVa-f there absorption maximums of variable intensity relating to vibrations of oxazolidine fragments [183, 213]: 1480-1465, 1445-1420 ("cutting" vibrations of CH_2), 1190-1175, 1160-1145, 1120-1105 (ν_{cycle}), 950-935 ("breathing" vibrations of cycle) and 850-830 cm^{-1}. In spectra of compounds CVIa-h there is a number of absorption bands of average intensity in the region 1170-1160, 1150-1100 and 1050-1035cm^{-1} characteristic for vibrations of imidazoline cycle [183].

In spectra of all synthesized heterocycles there absorption bands of variable intensity at 2700-2600, 1430-1410 and 800-670 cm^{-1} characteristic for vibrations of C–S bonds [167]. Together with mentioned vibrations in spectra of oxazolidines CIVc, 1 and CVf, and also imidazolines CVIc-h there is intensive narrow absorption band at 3640-3620cm^{-1} relating to stretching vibrations of HO-group in spatially hindered phenols [205].

In PMR spectra of oxazolidines CIVa-i signals of protons of methylene groups of cycle (4-CH_2 and 5-CH_2) are presented in the shape of "extended" multiplets in the region 3.86-5.54m.p. [183, 271], and in spectra of imidazolines CVIa-h – in the region 3.92-4.68m.p.. Signal of proton 5-H in spectra of 5-alkylthiomethyloxazolidines CVa-f is observed in the shape of non-symmetric multiplet in the interval 4.35-4.74m.p.. In spectra of 2-methyloxazolidines CIVa-d, j-m and CVa-f and also 2-methylimidazolidines CVIa-c, f-h signals of protons of methyl groups are revealed as broadened singlets at 1.54-1.70m.p. that is characteristic for methyl groups in cyclic systems [167, 183, 271].

The contents of individual components in mixtures of reaction products were determined for some of synthesized oxazolidines and imidazolines with the help of chromatography-mass spectrometry (Table 4.18).

Table 4.18. Individual composition of mixtures of synthesized heterocyclic compounds (in accordance with chromatography-mass spectrometry data)

Alk	Contents of components in compounds mixtures, %			
	CIVa	CIVi	CIVk	CVIIb
Et	2.9	2.8	3.2	3.0
Pr	22.5	21.6	21.5	20.8
i-Pr	41.3	40.5	41.0	40.2
Bu	5.6	5.5	5.5	5.7
i-Bu	22.2	23.6	22.4	24.1
n-C_5H_{11}	1.5	1.8	1.9	1.5
i-C_5H_{11}	4.0	4.2	4.5	4.7

4.4. Thiomethylation of Ketones as the Method of Demercaptanization of Sulfur-Containing Gas-Condensates, Oil and Sulfide-Alkaline Solutions

Removal of sulfur and especially of toxic and ecologically dangerous mercaptans from hydrocarbon raw materials is one of the most important problems in gas- and oil-refinery industries. Various methods of demercaptanization of oils and gas condensates are known, many of them due to some disadvantages (technological, apparatus, economic) are not realized in practice [110, 111, 116, 274, 275]. Methods of hydrocarbon raw materials purification applied in industry lead either to destruction of mercaptans to hydrogen sulfide and hydrocarbons (hydrorefining), or to transformation into disulfides (oxidative demercaptanization) having restricted application.

At present chemical neutralizers are offered for removal of hydrogen sulfide from oil and gas condensates, for example special reagent "Scavenger SX-2081" transform mercaptans into less volatile and toxic dialkylsulfides [276].

The data presented in Section 4.3.1 testify to high reactivity of natural C_2–C_5 mercaptans mixture in reaction of thioalkyltion of ketones. This fact underlies the research of possibility of demercapatanization of gas condensates, oil and their distillate fractions with simultaneous reception of concentrates of β-alkylthioketones by the interaction between ketones and mercaptans dissolved in hydrocarbon sulfur-containing raw materials.

Firstly we studied the interaction of acetone with formaldehydes and mercaptans obtained from Orenburg, Astrakhan and karachagan gas condensates, zhanazhol and tengiz oils, and also distillates isolated from them by the following reaction:

$$RSH + CH_2O + CH_3COCH_3 \xrightarrow[40\text{-}50^{\circ}C]{NaOH} RSCH_2CH_2COCH_3$$

Characteristics of hydrocarbon sulfur-containing raw materials used in our work are presented in Table 4.19. The data on composition of mercaptans presenting in listed gas condensates and oils were presented in Chapter 1.

Reactions of thiomethylation were carried out in the medium of sulfur-containing gas condensates, oil and their distillates at 20-50°C for 0.5-4 hours with the use of 20% water solution of sodium hydroxide and 35% solution of formaldehyde at molar ratio alkali : formaldehyde : acetone equal to 1 : 1 : 3 per 1g of mercaptane sulfur. The course of reaction was controlled by determination of content of mercaptan and sulfide sulfur by potentiometric titration (after 15 min) of samples of reaction medium by ammoniate solution of silver nitrate and potassium iodate accordingly [277]. The demercaptanization degree of initial raw materials (mercaptans conversion, K) was calculated by the formula:

$$\mathbf{K} = \frac{(\mathbf{S}^1_{\mathbf{merc.}} - \mathbf{S}^2_{\mathbf{merc.}}) \times 100}{\mathbf{S}^1_{\mathbf{merc.}}}$$

where $S^1_{merc.}$ and $S^2_{merc.}$ are the initial and final content of mercaptan sulfur, mass %.

Table 4.19. Characteristics of hydrocarbon sulfur-containing raw materials

Raw materials	Boil-off limits, °C	Density, ρ_4^{20}, g/cm^3	Content of sulfur, mass %			
			total, $S_{tot.}$	sulfide, $S_{sul.}$	mercaptan, $S_{merc.}$	"residual" [17]
Gas condensates						
Orenburglskoe field	40-292	0.695	1.07	0.26	0.58	0.23
Karachaganakskoe field	65-520	0.791	0.62	0.16	0.33	0.13
Astrakhan field	71-560	0.806	0.96	0.19	0.25	0.52
Oil						
Zhanazhol field	-	0.812	0.66	0.30	0.14	0.22
Tengiz field	-	0.805	0.60	0.14	0.05	0.41
Distillates of:						
zhanazhol oil	b.b.-180	0.704	0.43	0.19	0.215	0.025
	180-360	0.796	0.56	0.29	0.136	0.134
tengiz oil	b.b.-180	0.716	0.12	0.03	0.07	0.02
	180-350	0.788	0.62	0.10	0.06	0.46

We established that under intensive mixing of reaction mixture demercaptanization process proceeds fast enough (Figure 4.1). So, when used Astrakhan gas condensate for 5-10min at 20°C mercaptans conversion (K) reaches 70%.

Table 4.20. Change of structure of sulfur-containing components in hydrocarbon raw materials and conversion of mercaptans under acetone thiomethylation

Raw materials	Content of sulfur, mass %						Mercaptans conversion, %
	total		sulfide		mercaptan		
	initial	in reaction products	initial	in reaction products	initial	in reaction products	
Gas condensates:							
orenburg	1.07	1.07	0.26	0.68	0.58	0.009	95.0
karachaganaksk	0.62	0.62	0.11	0.25	0.33	0.007	97.9
astrakhan	0.96	0.96	0.18	0.28	0.25	0.011	99.6
Oil:							
zhanazholsk	0.66	0.66	0.30	0.41	0.14	0.007	95.0
tengizsk	0.60	0.60	0.14	0.17	0.05	0.003	94.0
Distillates of zhanazholsk oil							
b.b. – 180°C	0.43	0.43	0.19	0.42	0.215	0.012	94.4
180 – 360°C	0.56	0.56	0.29	0.42	0.136	0.015	89.0

The data on determination of mercaptan and sulfide sulfur before and after reaction (Table 4.20) testify to significant decrease of content of mercaptan sulfur at simultaneous increase of sulfide one and constant content of total sulfur. For example, in orenburg gas condensate the initial content of sulfide sulfur is 0.26 mass % (24.3% from S_{tot}) and after thiomethylation the part of $S_{sulf.}$ Is increased up to 0.68 mass % (63.5% from S_{tot}) where as $S_{merc.}$ is decreased from 54.2 down to 0.8% from S_{tot} (from 0.58 down to 0.009 mass %).

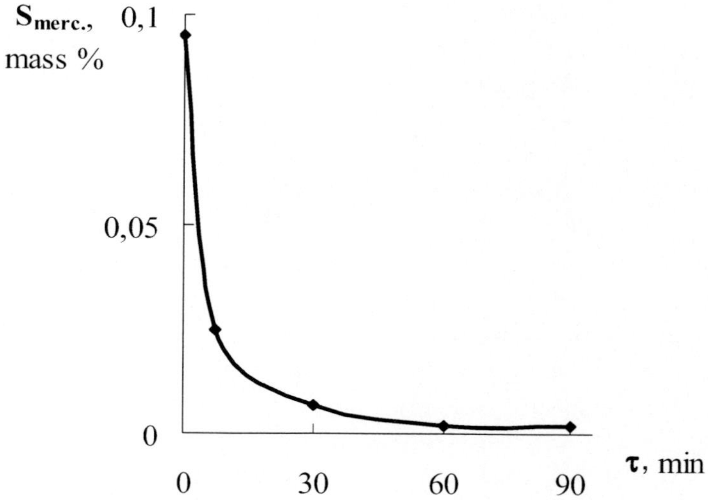

Figure 4.1. Decrease of $S_{merc.}$ under thiomethylation of propan-2-one by Astrakhan gas condensate.

At the same time the content of mercaptan sulfur in all samples of hydrocarbon raw materials after their thiomethylation does not exceed 0.001-0.015 mass % and degree of demercaptanization of sulfur-containing gas condensates, oil and distillates reaches 89.0-99.6%. Variation of demercaptanization conditions showed (Table 4.21) that conversion (K) of mercaptans and rate of thiomethylation of acetone are increased with the rise of $S_{merc.}$: CH_2O : acetone : NaOH from 1 : 1 : 0.5 : 0.5 up to 1 : 2 : 5 : 1 (moles) and rise of reaction temperature from 20 up to 50°C. So, at 1 : 1 : 0.5 : 0.5 (moles) and T = 20°C for 0.5 hour K reaches 68.7%, and for 3 hours – 84.1%, i.e. it is increased in 1.2 times. With the rise of amount of CH_2O and acetone up to 1 : 1.5 : 1 : 0.5 (moles)K = 88,2% (for 2 hours), and at 1 : 1.5 : 3 : 0.75 K = 94,2%.

Table 4.21. The influence of conditions of acetone thiomethylation by orenburg gas condensate on conversion of presenting in it thiols

Molar ratio of reagents				Reaction temperature, °C	Conversion $S_{merc.}$, %			
$S_{merc.}$	CH_2O	Acetone	NaOH		0.5h	1.0h	2.0h	3.0h
1	1.0	0.5	0.5	20	68.7	82.8	-	84.1
1	1.5	1.0	0.5	20	72.5	86.3	88.2	-
1	1.5	3.0	0.75	20	78.4	93.5	94.2	-
1	2.0	3.0	1.0	20	87.6	98.4	-	98.5
1	2.0	3.0	1.0	35	92.5	99.1	-	-
1	2.0	3.0	1.0	50	99.3	99.8	-	-

At double excess of CH_2O, trice excess of acetone and equimolar (in relation to $S_{merc.}$) amount of NaOH the value of K is increased up to 98.4% already after 1 hour at 20°C. Rise of reaction temperature up to 35°C insignificantly increases K (99.1% for 1 hour), and up to 50°C – up to almost complete conversion (99.3-99.8%) for 0,5-1 hour. Application of other than acetone ketones for thiomethylation of orenburg gas condensate at reagents ratio 1 : 2 : 3 : 1 (moles) and T = 20°C somewhat decreases conversion of mercaptans: in the case of methylethylketone – down to 92.3% for 1 hour, methylbutylketone – down to 85,4%, acetophenone – down to 77,5%. However at T = 50°C and reaction duration 3 hours even in the case of the least active acetophenone K = 95%. Change of water solution of alkali by aqueous-alcoholic (1 : 1) significantly increases rate of thiomethylation and mercaptans conversion, and decrease of NaOH amount (lower than 0.5 mole per 1g of mercaptan sulfur) acts in contrary direction.

Reaction may be carried out with both aqueous solution of CH_2O and paraform since it is easily depolymerized in alkaline medium. Application of paraform is ecologically advisable because odd water is removed out of reaction mixture. Application of paraform and ultrasonic irradiation decreases NaOH consumption.

Boiling temperature of formed under thiomethylation β-alkylthioketones is essentially higher than that one for initial mercaptans. For example *n*-butylmercaptan has boiling temperature 98.5°C and the product of its interaction with formaldehyde and methylethylketone (4-butylthio-3-methylbutanone-2) has boiling temperature equal to 255°C at 760 millimeter of mercury [278]. This leads to the change of distribution of sulfur-organic compounds in virgin distillates isolated from demercaptanized raw materials (Tables 4.22-4.24).

Table 4.22. Type content of sulfur-organic compounds of benzine fractions obtained on the base of hydrocarbon raw materials before the removal of mercaptans (A) and after demercaptanization (B)

Benzine fraction of	Content of sulfur, mass %.					
	Total		Sulfide		Mercaptan	
	A	B	A	B	A	B
zhanazhol oil (beginning of boiling – 180°C)	0.43	0.17	0.19	0.12	0.215	0.0012
orenburg gas condensate (beginning of boiling – 200°C)	0.71	0.31	0.26	0.29	0.400	0.001

It is obvious that demercaptanization of gas condensate and oil allows receiving benzine distillates which practically do not contain mercaptans and maybe transported or kept with out any limitations. In mentioned distillates the content of total sulfur is also low and vat residue represents concentrate of β-alkylthioketones which may find skillful practical application.

Table 4.23. The decrease of sulfur content in orenburg condensate after its demercaptanization by methylbutylketone

Fraction	Yield, mass %	Content of sulfur, mass %		
		total	Sulfide	Mercaptan
Initial gas condensate	100.0	1.07	0.26	0.58
Gas condensate after processing	100.0	1.07	0.68	0.009
beginning of boiling – 65°C	27.5	0.15	0.13	0.0
65–155°C	39.0	0.24	0.24	0.0

Table 4.24. The distribution of sulfur-organic compounds in benzine fraction of zhanazhol oil (beginning of boiling – 180°C) before and after its thiomethylation by methylbutylketone

Product	Yield, mass %	Content of sulfur, mass %		
		total	total	total
Initial benzine	100.0	0.43	0.19	0.215
Benzine	99.9	0.43	0.42	0.0012
Benzine strippant after distillation	37.8	0.17	0.12	0.001
Residual after distillation with boiling temperature higher than 180°C	1.9	11.8	11.6	-

Methylvinylketone was also used for demercaptanization of gas condensates. Methylvinylketone containing activated double bond is able to interact with mercaptans in the presence of bases (by thyilation reaction) forming 4-alkylthiobutanones-2 [10, 79]. At the same time its participation in mentioned reaction by CH_3–group under processing by CH_2O and mercaptans in alkali medium with formation of 1,5-di(alkylthio)pentanones-3 is also possible [248]. The last compounds may be obtained as a result of thyilation of intermediate reactive 5-alkylthio-1-pentenones-3.

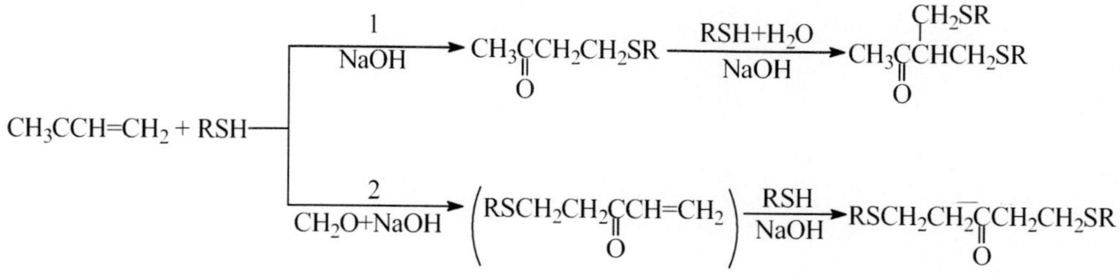

Demercaptanization of gas condensates was carried out at 20-50°C for 1-5 hours at molar ratio $S_{merc.}$: methylvinylketone : NaOH equal to 1 : 2 : 1 : 1 (thyilation reaction) or $S_{merc.}$: methylvinylketone : CH_2O : NaOH equal to 1 : 1-2 : 1-2 : 1 (thiomethylation reaction). In all cases we used 15% water solution of alkali. We established that processing of gas condensates by methylvinylketone by both variants is accompanied by decrease of mercaptan sulfur. Conversion of mercaptans in dependence on reaction conditions is varied between 72 and 92.7% (Figure 4.2). Conversion of mercaptans of orenburg gas condensate is higher than in the case of karachagan one, that is obviously conditioned by the presence in the first of them of significant amount of more reactive low-molecular mercaptans (Chapter 1) which hare dissolved in lighter hydrocarbon raw materials (boiling limits of orenburg gas condensate is 40–292°C, and of karachagan 65–520°C). Conversion of mercaptans is higher in the presence of formaldehyde (curves 1 and 2 in Figure 4.2) than in its absence (curve 3in Figure 4.2).

Demercaptanization process proceeds fast enough also at stoichiometric consumption of methylvinylketone: during 1 hour mercaptans conversion in orenburg gas condensate reaches 90%, and in karachagan one 69-73% that is insignificantly lower than maximum conversion at these conditions for 5 hours (Figure 4.2). Introduction of aqueous-alcohol solution of alkali (1 : 1) into the process instead of water one and also increase of consumption methylvinylketone and CH_2O up to 1.5-2.0 moles per 1g of mercaptan sulfur allows complete demercaptanizating of gas condensates. At that the content of mercaptan sulfur is decreased and content of sulfide sulfur is increased (Table 4.15) that indicates on formation of β-alkylthioketones and β,β'-di(alkylthio)ketones as products of thyilation and thiomethylation reactions.

These products of reactions are dissolved in gas condensates and in their IR spectra intensive absorption bands are revealed at 1705-1715cm^{-1} characteristic for stretching vibrations of carbonyl group in aliphatic ketones [167]. At the same time in gas condensates spectra after carrying out of thyilation and thiomethylation reactions there is no absorption in the regions 1680-1665 andи 1630-1620cm^{-1} typical for stretching vibrations of conjugated C=O and C=C-group and α, β-unsaturated ketones. This fact confirms formation of saturated ketosulfides or bisulfides in mentioned reactions.

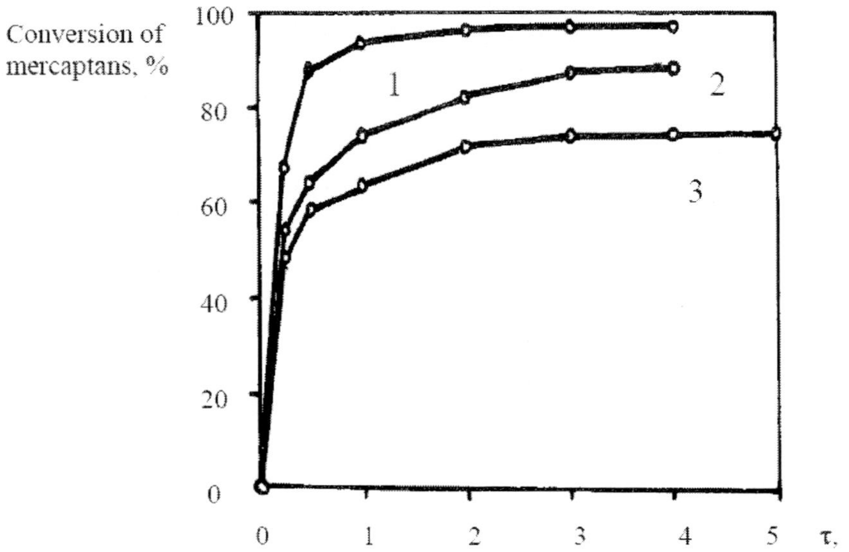

Figure 4.2. Conversion of mercaptans under thiomethylation (curves 1, 2) and thyilation (curve 3) of methylvinylketone by orenburg (1) and karachagan (2, 3)gas condensates.

Table 4.25. The change of type content of sulfur-organic compounds in gas condensates after their thiomethylation and thyilation by methylvinylketone

Gas condensates	Content of sulfur, mass %					
	$S_{merc.}$			$S_{sulf.}$		
	in raw materials	in thiomethylation products	in thyilation products	in raw materials	in thiomethylation products	in thyilation products
orendurg	0.58	0.008	0.008	0.26	0.65	0.66
karachagan	0.33	0.02	0.02	0.16	0.42	0.40

We mentioned above that vat residue (Table 4.24) of benzenes distillation underwent thiomethylation represented concentrate of β-alkylthioketones ($S_{tot.}$ = 11.8 mass %, $S_{sulf.}$ = 11.6 mass %). However their isolation out of gas condensates turned to be hard enough process. So, in presented work with this aim we used rectification and extraction of β-alkylthioketones by sulfuric acid or dimethylsulfoxide.

Vat residues (boiling temperature higher than 200°C, yield 20-35 mass %) containing not more than 3 mass % of sulfide sulfur (or not more than 25-35 mass % of β-alkylthioketones) were obtained by rectification of demercaptanized by acetone and formaldehyde orenburg and karachagan gas condensates. Consequently rectification method of ketosulfides isolation out of gas condensates is ineffective. So, we established that the best technique of β-alkylthioketones isolation out of demercaptanized hydrocarbon raw materials was extraction by 86% sulfuric acid at their volumetric ratio 1 : 5-10 (Table 4.26). Probably sulfuric acid extraction of β-alkylthioketones as in the case of sulfides [279] proceeds via formation of protonated complexes which are then easily destructed under hydrolysis.

Table 4.26. The influence of sulfuric acid concentration on degree of isolation of β-alkylthioketones by extraction out of orenburg gas condensate demercaptanized by acetone and formaldehyde

Concentration of sulfuric acid, mass %	60	70	80	85	86	87	88
The yield of ketosulfides concentrate, mass %	0.12	0.27	1.83	2.30	2.60	2.50	2.40

Concentrates of ketosulfides containing 9.5, 8.6 and 7.4 mass % of sulfide sulfur were obtained from preliminary demercaptanized orenburg and karachagan gas condensates and tengisk oil correspondingly by mentioned method. At the same time sulfuric acid isolation of β-alkylthioketones from karachagan gas condensate and tengisk oil is accompanied by strong resinification of extracts. Content of ketosulfides in isolated concentrates does not exceed 70-74% that is probably connected with co-extraction of other components of raw materials.

Concentrates of β-alkylthioketones isolated with the help of dimethylsulfoxide (or by watered dimethylsulfoxide containing 3-5 mass %of H_2O) from demercaptanized gas condensates contain 2.5-6.2 mass % of sulfide sulfur and not more than 60% of ketosulfides.

With the aim of reception of concentrates with higher content of β-alkylthioketones we studied the possibility of ketones thiomethylation by sulfide-alkaline solutions. In literature review (Chapter I) we noticed that under removing of hydrogen sulfide and mercaptans from wide fractions of hydrocarbons by sodium hydroxide large amount of non-recyclable sulfide-alkaline solutions containing C_1-C_3-mercaptides of sodium, sodium sulfide and hydroxide is formed. These solutions have odor nuisance, low thermal stability, high corrosion activity and toxicity. At present sulfide-alkaline solutions most often are drained to factory sewerage system with further injection into underground beds.

In given work were used sulfide-alkaline solutions as sulfur-containing reagents for thiomethylation of ketones. Sulfide-alkaline solutions were sampled from the first and second levels of treatment plant of propane-butane fraction of OGR and represent mixture of CH_3–, C_2H_5– and *iso*-C_3H_7-mercaptides of sodium, Na_2S and NaOH, water. Condensation of acetone with sodium sulfide and formaldehyde proceeds via stage of formation of corresponding sodium mercaptide (Na-derivative of 3-oxobutanthiole-1) that is experimentally confirmed by appearance on the curve of potentiometric titration of characteristic jump of potential in the region + 100 down to – 100mW (Figure 4.3).

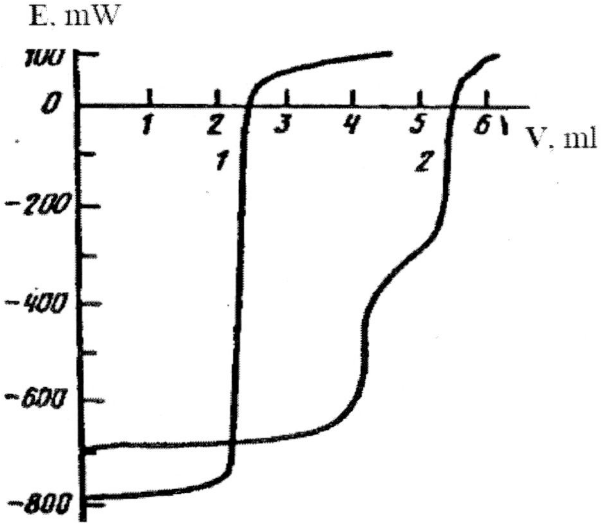

Figure 4.3. Curves of potentiometric titration of initial solution of sodium sulfide (1) and after addition of formalin and acetone into it (2).

$$Na_2S + CH_2O + H_2O \longrightarrow [NaS^- + \overset{\delta+}{C}H_2{=}\overset{\delta-}{O}] \longrightarrow NaSCH_2^+OH^-$$

$$CH_3COCH_3 \xrightarrow{OH^-} [CH_2COCH_2^-]$$

$$[CH_3COCH_2^-] + [NaSCH_2^+] \longrightarrow CH_3COCH_2CH_2SNa \xrightarrow[CH_3COCH_3]{CH_2O} (CH_3COCH_2CH_2)_2S + NaOH$$

Consequently under the use of acetone one of the products of sulfide-alkaline solutions thiomethylation is diketosulfide – di(3-oxobutyl)sulfide. At the same time sodium mercaptides presenting in sulfide-alkaline solutions usually give ketosulfides – 4-alkylthiobutanone-2.

$$RSNa + CH_2O \xrightarrow{H_2O} [\overset{\delta^+}{RS^-} + \overset{\delta^-}{CH_2{=}O}] \longrightarrow RSCH_2^+OH^-$$

$$[CH_3COCH_2^-] + [RSCH_2^+] \longrightarrow RSCH_2CH_2COCH_3$$

Table 4.27. Results of thiomethylation of ketones with the use of waste sulfide-alkaline solutions (temperature 20°C, reaction time 4 hours)

No. of experiment	Ketone	Ratio $S_{merc.}$: ketone : C H_2O, moles	Content of sulfur in reaction mixture, mass %		Conversion of sulfur-containing compounds, %	
			$S_{sulf.}$	$S_{merc.}$	$S_{sulf.}$	$S_{merc.}$
1	Acetone	1 : 2 : 1	1.70	3.60	82.9	80.8
2			2.08	3.37	75.0	75.3
3	Methylethy lketone	1 : 1 : 1	0.14	0.07	97.3	71.4
4			1.37	3.12	72.3	87.1
5			2.60	1.22	90.8	73.8
6	Acetophen one	1 : 1 : 1	1.70	3.60	76.4	96.5
7	Acetone	1 : 1,5 : 1	0.22	3.80	94.5	81.6
8		1 : 2 : 1	0.24	5.70	96.2	93.5
9	Methylethy lketone	1 : 0,5 : 0,5	0.22	3.80	92.3	75.0
10		1 : 1 : 1	0.22	3.80	100	96.6
11		1 : 1,5 : 1	0.22	3.80	100	98.7
12		1 : 1 : 1	0.49	2.50	77.6	90.0

Note: in experiments No. 1–6 – waste sulfide-alkaline solution from the first level of treatment of propane-butane fraction, in experiments No. 7–12 – from the second level of treatment.

It is obvious from Table 4.27 that thiomethylation of waste sulfide-alkaline solutions by ketones proceeds slow enough (2-6 hours) and rate of reaction and conversion of sulfur-containing compounds depend on both total amount of sulfide and mercaptan sulfur and ratio of sodium mercaptides and sulfides. Conversion of sodium sulfide is 77.6-100%, and of sodium mercaptides – 71.4-98.7%. The best results (experiments 8, 10 and 11) are reached at the ratio in initial raw material $S_{sulf.} : S_{merc.} \leq 0.06$ and use of methylvinyl ketone. Analysis of products showed that as a result of thiomethylation of ketones by mixture of CH_2O and waste sulfide-alkaline solutions the ketosulfides insoluble in water phase were predominantly formed. Characteristics of these ketosulfides (concentrates) containing more than 80-85% of β-alkylthio-ketones are presented in Table 4.28.

On the base of results of carried out investigations we developed principle technological scheme (Figure 4.4) of reception of concentrates of ketosulfides with simultaneous regeneration of waste sulfide-alkaline solutions. From reservoirs 1-3 sulfide-alkaline solutions, formalin and acetone are fed by pumps into reactor 4, products of their interaction are delaminated in collector 5 and are directed into reservoirs of desired products 6 (ketosulfides) and regenerated alkali 7.

Table 4.28. Characteristics of concentrates of β-alkylthioketones obtained on the base of waste sulfide-alkaline solutions from device of desulfurization of propane-butane fraction of OGR

Concentrate of β-alkylthioketone	n_D^{20}	d_4^{20}, g/cm^3	Content of sulfur, mass %			
			total	sulfide	disulfide	mercaptan
On the base of acetone	1.5639	1.1829	29.50	27.30	1.81	absent
On the base of methylethylketone	1.5636	1.1539	20.50	19.00	0.64	absent

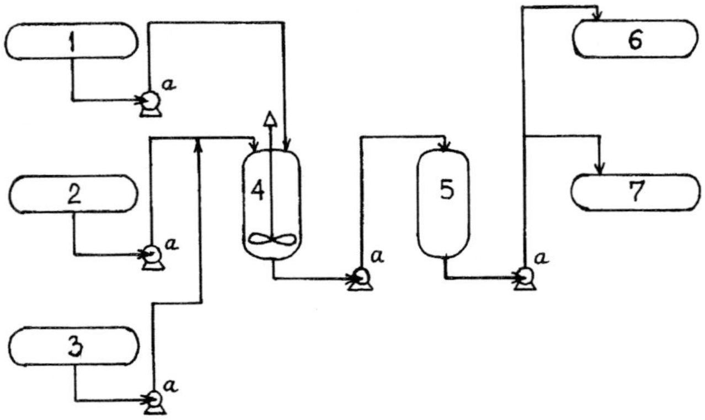

Figure 4.4. Principle technological scheme of regeneration of sulfide-alkaline solutions. 1-3 – reservoirs of sulfide-alkaline solution, formalin and ketone correspondingly, 4 – reactor, 5 – collector, 6 – reservoir of ketosulfides, 7 – reservoir of regenerated alkali, a – pumps.

Thus we showed that concentrates of β-alkylthioketones may be obtained as a result of thiomethylation of available ketones by various types of mercaptan-containing hydrocarbon raw materials and also by waste sulfide-alkaline solutions from desulfurization device of propane-butane fraction. At the same time reaction of thiomethylation of ketones may serve as effective method hydrocarbon raw materials demercaptanization which in particular allows decreasing the content of total sulfur in benzine fractions isolated under distillation of demercaptanized gas condensates.

4.5. Synthesis of β-Alkylthionitroalkanes on the Base of Mixture of C_2–C_5 Mercaptans

In literature there are some data [248, 280] on reactions of thioalkylation of primary nitroalkanes by interaction of the last ones with carbonyl compounds and aliphatic mercaptans in the presence of catalytic amount of bases. As a result β-alkylthionitroalkanes are formed with yields high enough:

$$RCH_2NO_2 + R^1COR^2 + R^3SH \xrightarrow{OH^-} R^1\text{-}\underset{SR^3}{\overset{R^2}{C}}\text{-}\overset{R}{HC}\text{-}NO_2$$

where R, R^1, R^3 = alkyl, R^2 = H, alkyl, aryl

We used this reaction for reception of β-alkylthionitroalkanes on the base of key compounds – mixture of natural mercaptans C_2–C_5. Such aliphatic nitro-compounds are of interest as semi-products in synthesis of β-aminosulfides [10] which can be hardly obtained by other methods.

We established by carried out experiments that under condensation of this mixture with nitromethane and aliphatic, aromatic and heterocyclic ketones the corresponding β-alkylthionitroalkanes CVIIa-j were formed. The best conditions of synthesis are: equimolar ratio of reagents, carrying out of reactions in ethanol (or propanol-2) solution in the presence of catalytic amounts of piperidine. By this method by boiling of C_2–C_5 mercaptans and aldehydes mixture for 6-10 hours the β-alkylthionitroalkanes CVIIa-f with the yield 63-76% were obtained (Table 4.29). If one uses in this reaction of less reactive ketones more prolonged boiling is necessary (15-18 hours) as a result of which corresponding β-alkylthionitroalkanes CVIIg-j with the yield 54-62% are formed.

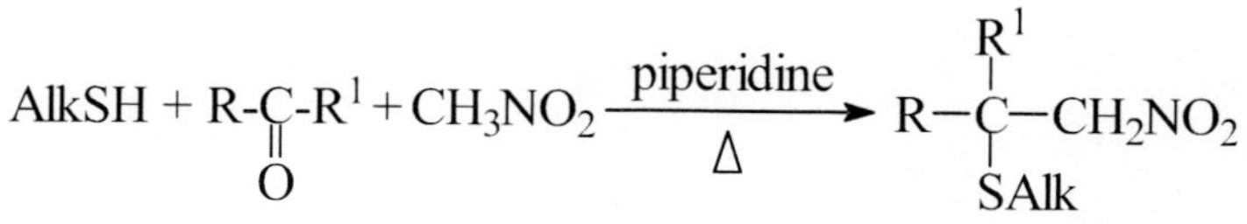

CVIIa-j

Alk = alkyl C_2-C_5;
R = C_2H_5 (a), *n*-C_3H_7 (b), Ph (c), 4-HO-3,5-(*tert*-C_4H_9)$_2$$C_6H_2$ (d), furyl-2 (e), pyridile-3 (f);
R^1 = H (a-f); R = R^1 = CH_3 (g); R = CH_3, R^1 = Ph (h), R = furyl-2, R^1 = CH_3 (i); R + R^1 = $(CH_2)_5$ (j)

Under the use of other catalysts (water solution of alkali, sodium ethylate, trimethylbenzylammonium hydroxide, potassium fluoride) the yield of adducts CVIIa-j is significantly decreased due to strong resinification of reaction products.

They suggest [10] that reaction proceeds with intermediate formation of nitroalkanes and further connection of mercaptans by activated double bond.

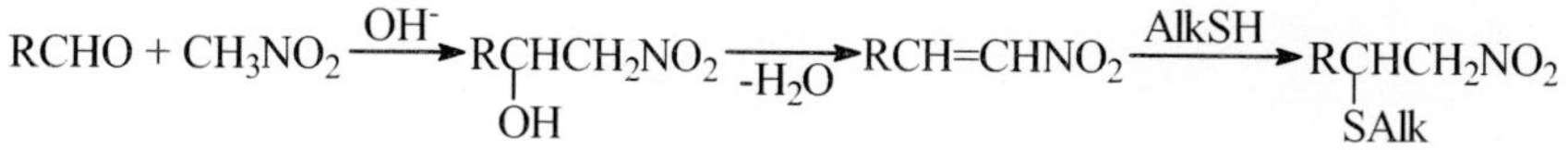

The confirmation of such reaction mechanism is the reception of 2-phenyl- CVIIc and 2-(furyl-2)-2-alkylthionitroethanes CVIIe by interaction of C_2–C_5 mercaptans fraction with corresponding β-nitroalkanes – β-nitrostyrene and β-(furyl-2)-α-nitroethylene in the presence of catalytic amounts of sodium methylate. Under stirring of equimolar amounts of reagents and sodium methylate at 30-35°C for 2-3 hours in absolute methanol the desired adducts CVIIc, e were obtained with the yield 74-78%.

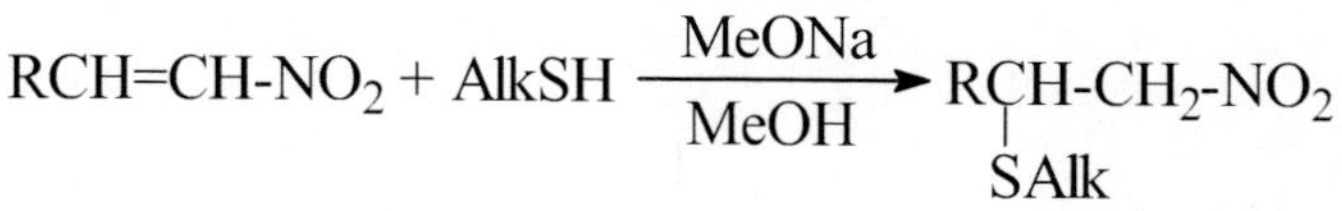

CVIIc, e

Alk = alkyl C_2-C_5; R = Ph (**c**), furyl-2 (e)

Table 4.29. Constants, yield and data of elementary analysis of β-alkylthionitroalkanes $AlkSCRR^1CH_2NO_2$

Compound	R	R^1	Yield, %	$T_{boil.}$, °C (millimeters of mercury) or $T_{melt.}$, °C (solvent for crystallization)	n_D^{20}	d_4^{20}, g/cm³	S, mass %
CVIIa	C_2H_5-	H-	68	80-101 (2)	1,5340	0,9613	17,71
CVIIb	*n*-C_3H_7-	H-	65	93-114 (2)	1,5288	0,9635	16,42
CVIIc	Ph-	H-	70	54-59 (ethanol-water, 2:1)	-	-	13,48
CVIId	4-HO-3,5-(*tert*-$C_4H_9)_2C_6H_2$-	H-	75	47-52 (benzene-hexane, 4:1)	-	-	9,04
CVIIe	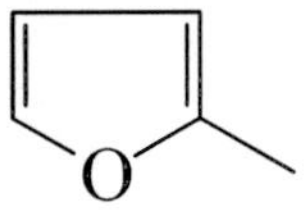	H-	63	127-145 (1)	1,5402	1,0012	14,04
CVIIf	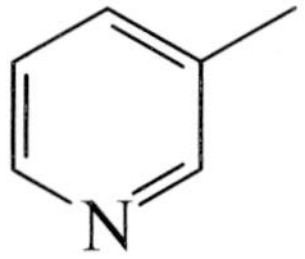	H-	76	84-90 (propanol-2-water, 1:1)	-	-	13,75
CVIIg	CH_3-	CH_3-	62	108-128 (1)	1,5256	0,9601	17,58
CVIIh	Ph-	CH_3-	58	62-68 (толуол)	-	-	13,32
CVIIi	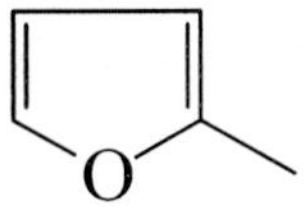	CH_3-	54	42-47*)	-	-	14,05
CVIIj	-$(CH_2)_5$-		56	oil*)	1,5419	0,9684	14,77

Note: *) The compounds were purified by column chromatography on silica gel.

Physical-chemical characteristics of synthesized β-alkylthionitroalkanes are presented in Table 4.29. Structures of these compounds are confirmed by IR and PMR spectra data. So, in IR spectra of mentioned compounds there are intensive absorption maximums in the region 1555-1545 and 1370-1360cm^{-1} characteristic for asymmetric and symmetric stretching vibrations of nitro-groups in aliphatic compounds and also the number of bands of variable intensity at 2700-2600, 1430-1420 and 790-695cm^{-1} relating to vibrations of bonds C–S [166, 167, 209].

In PMR spectra protons of nitromethylene group are revealed as doublets (compounds CVIIa-f) or singlets (compounds CVIIg-j) in the region 4.05-4.17m.p. that is characteristic for aliphatic nitro-compounds [167].

Chapter 5

Testing of Applied Properties of Organic Compounds Synthesized on the Base of C_2–C_5 Mercaptans Mixture

As we have mentioned in Chapter 1 many sulfur-organic compounds synthesized on the base of individual thioles and in some cases technical mixtures of thioles possess valuable applied properties and may be applied as additives improving anti-oxidative, anticorrosion, antiwear and other properties of lubricating oils, polymer compositions and also as biologically active substances.

In this connection we tested efficiency of some compounds synthesized on the base of thioles C_2–C_5 mixture and available chemical and petrochemical products in mentioned fields of application.

5.1. Thioalkyl Ethers of Carbonic Acids – Mineral Lubricating Oils Additives

We tested derivatives of thiocarbonic acids synthesis of which was described in Section 3.2. They were dissolved in mineral oil M-11 used for preparation of many industrial lubricating compositions [72] and thermal oxidative stability, anticorrosion and antiwelding properties of obtained compositions were estimated by known techniques [284]. Results of tests are presented in Table 5.1.

Results of tests showed that among synthesized S-alkyl ethers of thiocarbonic acids there were compounds with good anticorrosion properties. For example under addition into M-11 of 2 mass % of ethers of acids LXIXg, LXIXh and LXIXj corrosion is decreased from 188.5 down to 4.5-6.7 g/m^2; in the presence in oil of the same amount of industrial additives DF-11 and IKhP-21 corrosion was decreased only down to 24.6 and 18.4 g/m^2. Ether LXIXf is inferior to them. Ethers of alkane thionic acids LXIXa-c, benzene carbothionic LXIXd and 1-naphthalene carbothionic LXIXe acids reveal the lowest anticorrosion activity.

In given regime of friction some ethers are antiwear and antiwelding additives effective enough. Ethers LXIXh, LXIXi and LXIXj reveal the best antiwear properties; at their concentration in M-11 2 mass % diameter of wear scar (D_w) is decreased from 0.89 down to

0.35-0.40mm, and tearing index (I_t) is increased from 24 up to 50-55. Ethers LXIXc and LXIXg are also effective. Mentioned compounds exceed in antiwear and antiwelding action additives DF-11 and IKhP-21 especially at work concentrations 1-2 mass % as characteristic for additives of such purpose [72].

Table 5.1. Results of tests of S-alkyl ethers of thiocarbonic acids as additives for mineral oil M-11

Content of additive, mass %	Corrosion, g/m^2	Thermooxidative stability at 250°C, min	Antiwear properties	
			D_w, mm	I_t
With additive LXIXa				
0.5	104.7	30	0.84	25
1.0	88.5	32	0.78	27
1.5	72.4	36	0.70	30
2.0	63.0	38	0.66	31
With additive LXIXb				
0.5	128.6	28	0.82	28
1.0	85.3	31	0.77	30
1.5	66.5	34	0.71	31
2.0	44.4	36	0.67	33
With additive LXIXc				
0.5	115.8	27	0.80	27
1.0	77.5	29	0.72	30
1.5	60.8	32	0.65	33
2.0	48.5	34	0.54	36
With additive LXIXd				
0.5	136.6	30	0.77	28
1.0	100.5	32	0.72	29
1.5	84.4	35	0.70	34
2.0	61.6	38	0.64	36
With additive LXIXe				
0.5	96.6	26	0.81	27
1.0	74.5	31	0.70	30
1.5	50.8	34	0.64	34
2.0	37.2	36	0.58	38
With additive LXIXf				
0.5	66.6	40	0.70	32
1.0	54.3	44	0.64	34
1.5	25.4	52	0.59	35
2.0	14.8	58	0.52	38
With additive LXIXg				
0.5	32.4	36	0.69	36
1.0	9.9	40	0.60	38
1.5	8.0	42	0.52	42
2.0	6.7	47	0.47	44
With additive LXIXh				
0.5	26.6	43	0.58	42
1.0	15.3	51	0.50	44
1.5	9.4	57	0.43	47
2.0	4.5	65	0.37	50

Table 5.1. Continued

Content of additive, mass %	Corrosion, g/m^2	Thermooxidative stability at 250°C, min	Antiwear properties	
			D_w, mm	I_t
With additive LXIXi				
0.5	78.9	38	0.62	41
1.0	65.0	44	0.54	45
1.5	41.8	47	0.44	49
2.0	20.4	51	0.35	52
With additive LXIXj				
0.5	53.5	48	0.64	40
1.0	26.3	51	0.52	46
1.5	13.6	54	0.47	50
2.0	5.4	60	0.40	55
With additive DF-11				
0.5	74.3	33	0.82	26
1.0	50.2	37	0.73	30
1.5	33.7	40	0.59	31
2.0	24.6	45	0.48	33
With additive IKhP-21				
0.5	62.4	38	0.76	24
1.0	46.3	44	0.72	22
1.5	24.2	47	0.68	24
2.0	18.4	54	0.64	27
Oil M-11				
-	188.5	18	0.89	24

Some studied ethers (LXIXf, LXIXh, LXIXj) increase thermal oxidative stability of oil M-11 (TOS_{250}) in 1.2-1.3 times. Compounds LXIXg and LXIXi are inferior to these ethers but they exceed DF-11 and IKhP-21. for the rest studied ethers index TOC_{250} is on the level of these industrial additives.

Thus, the results of investigations reported in this section of research testify to advisability of reception and application of S-alkyl ethers of thiocarbonic acids on the base of C_2–C_5-fraction of natural mercaptans as polyfunctional additives for mineral lubricating oils. Ethers including fragments of N-containing heterocycles indole, pyridine and benzthiazole most effectively improve anticorrosion, antiwear and antiwelding properties and also thermooxidative ability of base oil M-11.

5.2. N- and P-Containing Derivatives of 1-Alkylthiopropanol-2 – Mineral Lubricating Oils Additives

Mentioned N- and P-derivatives (LIc, d, h-j, LVII-LIX) were synthesized by known reactions presented in Section 3.1.2. Synthesized compounds are in the majority structure analogous of alkylsubstituted phosphates, diethyldithiocarbamates and dithiophosphates, known as additives for lubricating oils and polymer materials [8, 50, 72]; but in contrast to them they contain C_2–C_5-thiopropanol fragments as substituents. During tests this fact allows establishing the influence of given substituents in molecular structure of one-type products on efficiency of their functional action.

That is why as objects of comparison we chose polyfunctional additives DF-11 and INKhP-21 of dithiophosphate type [72] which active substance is also used as thermostabilizer of synthetic polymers [8] and organic accelerators of rubbers vulcanization [50]. Results of comparable tests of anticorrosion, antiwear and antiwelding properties and also thermal oxidative stability of mineral base oil M-11 containing 2 mass % of known additives and those ones synthesized by us are presented in Table 5.2.

Results of tests of mineral oil M-11 containing 2 mass % of ethers LIc, d, h-j showed good anticorrosion properties (Table 5.2). For example corrosion in the case of ethers LIc, d, LVIII is decreased from 188.5 down to 2.0-5.6g/m^2, and for DF-11 and IKhP-21 only down to 24.6 and 18.4g/m^2. By anticorrosion action ether of phosphorous acid LVII is inferior to these compounds. Ethers LIi and LIj are less active and decrease corrosion down to 27.2 and 31.4g/m^2 correspondingly.

Ethers LVII and LIXa, b reveal the best anticorrosion properties: diameter of wear scar is decreased from 0.89 down to 0.30-0.33mm, tearing index is increased from 24 up to 54-58. Ethers LIc, d are close to them. In the whole, by antiwear and antiwelding properties synthesized ethers are better than additives DF-11 and IKhP-21, for example change of D_w for them with the rise of their concentration in oil from 0.5 up to 2 mass % occurs at higher level and in more wide diapason (0.48-0.82mm) than for compounds LVII and LIXa, b (0.30-0.58mm).

Moreover tested ethers increase thermal oxidative stability of base oil and by this index (TOS_{250}, min) compounds LIc (70), LIh (75), LVIII (72) are inferior to DF-11 and IKhP-21 (45 and 54). The same regularity is observed also at 0.5% concentration of mentioned compounds: TOS_{250} is varied in the interval 37-54min, for DF-11 – 33, IKhP-21 – 38 minutes. The rest ethers are close to the last additives (LIi, LIj and LIXa, b).

Table 5.2. Results of testing of N- and P-containing derivatives of 1-alkylthiopropanol-2 as additives (2.0 mass %) for mineral oil M-11

Compound	Corrosion, g/m^2	Thermal oxidative stability of M-11 at 250°C, min	Antiwear and antiwelding properties	
			D_w	I_t
LIc	3.7	70	0.41	51
LId	2.0	56	0.37	50
LIi	27.2	52	0.62	40
LIj	31.4	57	0.51	37
LIh	16.6	75	0.54	36
LVII	14.2	43	0.30	58
LVIII	5.6	72	0.48	47
LIXa	21.3	60	0.38	54
LIXb	22.8	61	0.33	57
DF-11	24.6	45	0.48	33
IKhP-21	18.4	54	0.64	27
Oil M-11	188.5	18	0.89	24

Thus, N- and P-derivatives of 1-alkylthiopropanol-2 obtained on the base of natural mercaptans C_2–C_5 are able to increase anticorrosion, antiwear and antiwelding properties and also thermal oxidative stability of mineral oil up to higher level in comparison with industrial P- and S-containing additives DF-11 and IKhP-21.

5.3. Heterocyclic Sulfides – Additives for Mineral and Synthetic Lubricating Oils

In one of the previous sections of book (Section 2.4) we mentioned that among compounds of *sym*-triazine type effective antioxidants, thermo- and light stabilizers of polymers materials, lubricating oils and hydrocarbon fuels differing in high chemical and thermal stability were already found [142]. Presence of three atoms of N alternating with carbon atoms in cyclic molecule imparts to *sym*-triazine high adsorption (and simultaneously passivating) activity in relation to metals, and functional groups of substituents introduced into cycle are able to impart anti-oxidative and protecting properties. Sulfur atoms in these substituents may significantly strengthen mentioned properties of derivatives of *sym*-triazine [142].

Protecting properties of sulfides of *sym*-triazines, synthesis and properties of which are described in sections 2.3 (Table 2.26) and 2.4 (Table 2.28 and 2.29) were tested in mineral oils – industrial I-40 and motor M-11 using known techniques (GOST 9.054–75, methods 1, 4 and 5). Tests by method 1 were carried out in thermal moisture-chamber G-4 in regime of periodic condensation of moisture on samples St 20. Then samples were visually examined every equal time intervals from the beginning of tests until appearance of the first corrosion center. Tests by method 4 were carried out at constant embedding of samples into electrolyte (pH=8), by method 5 – under influence on sample by 0.1% water solution of HBr-acid. The object for comparison was known corrosion inhibitor Penazolin 10-16 (mixture of 1,2-disubstituted imidazolines – product of condensation of C_{10}–C_{16} with polyethylenepolyamines [132]). Results of tests are presented in Table 5.3.

All studied derivatives of *sym*-triazine reveal anticorrosion effect especially with the rise of concentration from 0.5 up to 1 mass % in both oils I-40 and M-11.

Sulfides XXXVc, XXXVIIc and XXXVe containing *n*-octadecyl, methyl and *n*- toluidine substituents and also XXXVj and XXXVIIg with one or two morpholine substituents possess the best protecting properties. So, in the presence of 1 mass % of sulfide XXXVc in oil I-40 the first sites of corrosion on samples are appeared after 45 hours, and in oil M-11 – after 50 hours, in the presence of sulfide XXXVe correspondingly after 42 and 40 hours (corrosion of steel was 5%).

Derivatives of *sym*-triazine containing fragments of diethylamine (XXXVm, XXXVIIi), octylamine (XXXVb, XXXVIIb) and spatially-hindered phenol (XXXVg) by efficiency of protecting action in oils I-40 and M-11 are inferior to other tested sulfides although exceed inhibitor Penazolin 10-16. Sulfides of *sym*-triazine with octadecyl substituents (XXXVc, XXXVIIc) reveal better protecting properties than their analogous with octyl substituents (XXXVb, XXXVIIb). Most probably the presence in molecular of additives of longer alkyl substituents provides their better solubility in oils, high energy of intermolecular cohesion and high stability of formed colloidal systems promoting by that formation of strong adsorption films on the surface of metal protected by them.

Table 5.3. Results of tests of triazine sulfides as oil-soluble corrosion inhibitors St 20

Oil	Method 1 (in chamber G-4)			Method 4 (in electrolyte solution)		Method 5 (in solution of HBr)	
	Time before corrosion site is appeared, h	Number of corrosion sites	Corrosion degree, %	Number of corrosion sites	Corrosion degree, %	Number of corrosion sites	Corrosion degree, %
With sulfide XXXVb							
I-40	21/25	54/30	26.5/15	35/12	17/6	9/5	5/0
M-11	23/18	68/38	34/18	30/15	15/7	14/6	7/3
With sulfide XXXVc							
I-40	38/45	18/0	9/0	2.5/0	12/0	10/0	5/0
M-11	34/50	40/0	20/0	0/0	0/0	0/0	0/0
With sulfide XXXVg							
I-40	25/20	42/38	21/19	20/18	10/9	30/18	15/9
M-11	27/19	58/34	19/17	40/14	20/7	21/8	10.5/4
With sulfide XXXVj							
I-40	33/38	30/25	15/12.5	18/15	9/7.5	35/12	17/6
M-11	32/36	40/0	20/0	0/0	0/0	30/5	15/3.5
With sulfide XXXVm							
I-40	24/30**)	52/18	26/9	40/14	20/7	10/0	5/0
M-11	22/31	47/25	24/12	32/10	16/5	12/8	6/4
With sulfide XXXVIIb							
I-40	21/24	54/42	27/21	42/18	21/9	15/8	7/4
M-11	18/22	75/38	36/19	30/15	14/7	21/10	11/5
With sulfide XXXVIIc							
I-40	36/39	30/18	15/8.5	18/15	9/7.5	35/12	17/6
M-11	33/37	40/0	20/0	0/0	0/0	30/0	14.5/0
With sulfide XXXVe							
I-40	39/42	47/0	23/0	25/0	12/0	12/0	6/0
M-11	32/40	35/10	18.5/5	0/0	0/0	0/0	0/0
With sulfide XXXVIIg							
I-40	30/33	47/0	23/0	25/0	12/0	11/0	6/0
M-11	32/35	40/30	20/15	0/0	0/0	0/0	0/0

Table 5.3. Continued

Oil	Method 1 (in chamber G-4)			Method 4 (in electrolyte solution)		Method 5 (in solution of HBr)	
	Time before corrosion site is appeared, h	Number of corrosion sites	Corrosion degree, %	Number of corrosion sites	Corrosion degree, %	Number of corrosion sites	Corrosion degree, %
With sulfide XXXVIIi							
Oil	Method 1 (in chamber G-4)			Method 4 (in electrolyte solution)		Method 5 (in solution of HBr)	
	Time before corrosion site is appeared, h	Number of corrosion sites	Corrosion degree, %	Number of corrosion sites	Corrosion degree, %	Number of corrosion sites	Corrosion degree, %
I-40	18/20	42/38	21/19	20/18	10/9	30/18	15/9
M-11	17/19	58/34	29/17	40/14	20/7	21/8	10.5/4
With inhibitor Penazolin 10-16							
I-40	22/23	54/50	27/25	CC***/40	CC /20	44/32	22/16
M-11	17/22	85/36	42/18	CC/10	CC /5	10/0	5/0
Oils without inhibitor							
I-40	15	62	31	CC	CC	81	40
M-11	14	94	47	CC	CC	34	17

Notes: *)Number of *sym*-triazine sulfides are presented in Tables 2.28 and 2.29;

**)In numerator there are data for 0.5% content of sulfide, in denominator – for its 1% content;

***) CC – change of colour.

Table 5.4. Results of testing of sulfides of *sym*-triazine as polyfunctional additives to mineral base oil M-11

Content in oil, mass %	Corrosion, g/m^2	Thermal oxidative stability at 250°C, min	Antiwear and antiwelding properties	
			D_w	I_h
With sulfide XXXVa				
0.5	57.5	26	0.62	40
1.0	31.2	29	0.56	43
1.5	17.1	32	0.48	45
2.0	8.9	36	0.41	47
With sulfide XXXVb				
0.5	17.2	38	0.52	43
1.0	10.3	43	0.44	47
1.5	5.2	48	0.42	53
2.0	2.7	52	0.36	55
With sulfide XXXVc				
0.5	11.8	39	0.56	46
1.0	5.9	42	0.46	52
1.5	4.3	47	0.41	54
2.0	1.2	54	0.34	57
With sulfide XXXVh				
0.5	15.2	43	0.73	25
1.0	7.2	51	0.64	27
1.5	5.0	57	0.55	29
2.0	3.8	65	0.50	33
With sulfide XXXVi				
0.5	60.5	35	0.49	44
1.0	40.2	40	0.43	47
1.5	21.7	43	0.38	50
2.0	9.4	47	0.34	54
With sulfide XXXVj				
0.5	28.5	45	0.58	31
1.0	12.4	53	0.50	33
1.5	9.3	60	0.44	37
2.0	7.0	71	0.40	41

Table 5.4. Continued

Content in oil, mass %	Corrosion, g/m^2	Thermal oxidative stability at 250°C, min	Antiwear and antiwelding properties	
			D_w	I_h
With sulfide XXXVk				
0.5	92.2	30	0.67	30
1.0	57.5	34	0.55	32
1.5	22.7	37	0.50	40
2.0	14.6	40	0.47	43
With sulfide XXXVIIa				
0.5	75.7	25	0.83	34
1.0	42.6	33	0.76	37
1.5	34.2	35	0.65	41
2.0	26.6	41	0.62	44
With sulfide XXXVIIb				
0.5	69.6	31	0.77	28
1.0	50.1	37	0.62	30
1.5	37.3	40	0.58	35
2.0	25.4	43	0.52	39
With sulfide XXXVIIc				
0.5	20.5	28	0.61	31
1.0	8.8	32	0.53	34
1.5	7.4	35	0.47	37
2.0	6.1	36	0.43	40
With sulfide XXXVIIg				
0.5	88.5	30	0.80	31
1.0	37.8	33	0.74	34
1.5	33.5	37	0.65	39
2.0	28.2	39	0.55	42
With sulfide XXXVIIh				
0.5	81.8	48	0.77	29
1.0	50.2	51	0.71	32
1.5	31.5	54	0.62	35
2.0	10.2	60	0.58	37

Table 5.4. (Continued)

Content in oil, mass %	Corrosion, g/m^2	Thermal oxidative stability at 250°C, min	Antiwear and antiwelding properties	
			D_w	I_h
With additive DF-11				
0.5	74.5	33	0.82	26
1.0	50.2	37	0.73	30
1.5	33.7	40	0.59	31
2.0	24.6	45	0.48	33
With additive IKhP-21				
0.5	62.4	38	0.75	24
1.0	46.3	44	0.72	22
1.5	24.2	47	0.68	24
2.0	18.4	54	0.64	27
Oil M-11 without additives				
-	188.5	18	0.89	24

Thus, N-substituted 2,4-diamino-6-alkylthio- and 2-amino-4,6-dialkylthio-*sym*-triazines synthesized on the base of C_2–C_5 natural mercaptans and containing higher alkyl substituents of normal structure, fragments of arylamine or morpholine reveal high protecting effect and may be applied as corrosion inhibitors in industrial and motor oils.

Such series of sulfides of *sym*-triazines which part has somewhat other structure of substituents was tested as polyfunctional additives for mineral base oil M-11 (in comparison with industrial additives DF-11 and IKhP-21) by methods analogous to those ones reported in Section 5.1. Results of tests presented in Table 5.4 show that all studied N-substituted 2,4-diamino-6-alkylthio-*sym*-triazines especially XXXVb, XXXVc and XXXXd are characterized by high anticorrosion properties. So, at 2% content of mentioned compounds in oil corrosion is decreased from 188.5 down to 1.2-3.8g/m^2. Sulfides XXXVa, XXXVj, XXXVk, XXXVIIi and XXXXIVb are somewhat inferior in this to more active compounds but also exceed additives DF-11 and IKhP-21.

The majority of tested 2-amino-4,6-dialkylthio-*sym*-triazines (XXXVIIa, XXXVIIb, XXXVIIh, XXXVIIg) are on the same level of anticorrosion action as known additives obviously due to reduced polarity and adsorption ability on plumbum of their molecules containing only one amino-group. The only exception in this raw is high active sulfide XXXVIIc which molecules contain substituent n-$C_{18}H_{37}$ as the most effective XXXVc reducing corrosion down to 4.3g/m^2 even at 1% content in oil.

Sulfides XXXVb, XXXVc and XXXVi possess the best antiwear properties – at their 2% content in oil the value of D_w is decreased from 0.89 down to 0.34-0.36mm and I_t is increased from 24 up to 54-57. Higher antiwear action in comparison with additives DF-11 and IKhP-21 are also revealed by compounds XXXVa, XXXVk and XXXVIIc, i. The rest derivatives of *sym*-triazine (XXXVh, XXXVj, XXXVIIb, XXXVIIg) in their antiwear and antiwelding properties are close to mentioned additives for which for example change of D_w with the rise of concentration in M-11 from 0.5 up to 2% occurs at higher level and in more wide diapason (0.48-0.82mm).

Synthesized derivatives of *sym*-triazine especially XXXVh, XXXVk and XXXVIIh in 3.3-3.9 times increase thermal oxidative stability of oil (Table 5.4). Sulfides XXXVb, XXXVc, XXXVi and XXXVIIb by the influence on TOS_{250} index ate close to industrial additives taken for comparison.

Compositions of various antioxidants strengthening efficiency of action of each other at the expense of synergism are used for improvement of antioxidative and anticorrosion properties of mineral oils [72]. For example for stabilization of turbine oil paired antioxidative additives *p*-oxydiphenylamine–dithioaniline and N-phenyl-β-naphthylamine (Neozon D)–anthranylic acid [281], zinc dithiophosphate (DF-11)–mixture of aliphatic amines C_{17}–C_{21} [282], etc. are recommended. Earlier they showed [283] that effective stabilization of industrial oils against oxidation might be reached with the help of compositions of additive Ionol (2,6-di-*tert*-butyl-4-methylphenol) with aminoderivatives of *sym*-triazine containing higher alkyl radicals.

In given work we studied the influence of Ionol composition with sulfides of *sym*-triazine and other heterocyclic compounds (derivatives of C_2–C_5 mercaptans mixture) on thermal oxidative stability of one of the types of industrial oil – bobbin oil of brand AU (GOST 1642-75). This oil is obtained from waxy oil after their advanced sulfuric acid purification; it is applied as work liquid in hydrosystems of various machines and devices and also as base for production of special work liquids and some lubricating oils [284].

As mentioned sulfides – components in compositional additives we also used described in Section 2.3 (XXXVg, h, Table 2.28; XXXXc, d, XXXXIVa, Table 2.31); 3.1 (LIh, Table 3.3; LXe, r, u, Table 3.9); 4.1 (LXXXIk, n, p, Table 4.1); 4.2 (LXXXVIb, e, Table 4.4) heterocyclic compounds. Compositions of additives included listed sulfides and Ionol at ratio 1 : 15, additives were introduced into oil AU with concentration 1 mass %. Thermal oxidative stability of oil with additives was estimated (Table 5.5) every 30 minutes in the process of oxidation by air oxygen in microreactor (150°C,1.5 hour) by the change of optical density (D) which was determined with the help of photoelectrical colorimeter PhEC-56M in cuvette with thickness 2mm [285].

Table 5.5. Change of optical density D of bobbin oil AU inhibited by compositional additives (ionol : sulfides = 15 : 1) under its oxidation (150°C, [InH] = 1 mass %)

Oil with additives:	Value of D before the beginning of oxidation	Value of D during the oxidation, h		
		0.5	1.0	1.5
Ionol (I)	0.054	0.151	0.204	0.213
I + XXXVg	0.065	0.084	0.133	0.149
I + XXXVh	0.057	0.091	0.118	0.208
I + XXXXc	0.055	0.062	0.083	0.098
I + XXXXd	0.050	0.086	0.117	0.192
I + XXXXa	0.043	0.082	0.107	0.114
I + LIh	0.040	0.102	0.129	0.146
I + LXXXIk	0.075	0.116	0.121	0.128
I + LXXXIn	0.063	0.163	0.225	0.264
I + LXXXIp	0.049	0.090	0.111	0.124
I + LXXXIIe	0.047	0.075	0.102	0.130
I + LXXXVIb	0.069	0.113	0.167	0.172
I + LXXXVIe	0.060	0.159	0.217	0.256
I + LXe	0.060	0.142	0.179	0.195
I + LXr	0.050	0.073	0.091	0.103
I + LXu	0.052	0.111	0.154	0.184
oil without additives	0.046	0.227	0.289	0.313

As it is obvious from data of Table 5.5 under oxidation of oil without additives the value of D is changed significantly reaching the value 0,314 at the end of experiment. Addition of 1 mass % of Ionol increases TOS of oil and decrease of value of D down to 0.213 indicates on it. Introduction into oil of Ionol compositions with testing sulfides strengthen inhibiting effect some more: in the majority of cases optical density of oxidized oil is decreased in 2.5 times.

Analysis of results of tests of bobbin oil AU with compositional additives allows disposing all studied second components (sulfides) in the order of decreasing of their efficiency in the following raw:

XXXXc > LXr ~ XXXXIVa > LXXXIp > LXXXIk > LIh > LXXXIIe > XXXVg > LXXXVIb > LXu > XXXXd > LXe > XXXVh > ionol > LXXXVIe > LXXXIn

Compositions of Ionol with derivatives of nitrogen-containing heterocycles containing in two fragments of phenol spatially hindered by substituted oxazolidines LXr, *sym*-triazines XXXXc and XXXXIVa, and also composition of Ionol with LXXXIp including fragment of 2,2,6,6-tetramethylpyperidine reveal the highest inhibiting activity in 1.7-2.2 times higher in comparison with industrial antioxidant. At the same time compositions containing derivatives of benzthiazole (compounds LXXXIn and LXXXVIe) are inferior in efficiency to Ionol action.

High efficiency of action of compositional additive (Ionol and XXXXc) allows decreasing its sum concentration in oil. For example at concentration 0.5 and 0.25 mass % the additive keeps practically the same inhibiting activity (Figure 5.1) that at concentration 1 mass % (Table 5.5). Even under decrease of concentration down to 0.05 mass % this compositional additive maintains its antioxidative stability of oil at higher level in comparison with oil containing 1 mass % of Ionol.

We also studied the influence of ratio of Ionol to second component on efficiency of action of compositional additives on the example of its mixture XXXXc, XXXXd, LXXXIp and LXXXIIe. These compounds are differed in structure but contain functional groups responsible for antioxidative properties of compound in the whole: fragments of spatially-hindered phenol, cyclic screened amine or diphenylamine. Results of these investigations are presented in Table 5.6 and in Figure 5.2 and it is obvious from them that ratio Ionol : second component essentially influences on efficiency of composition and its change may lead to both synergism and antagonism of additives action.

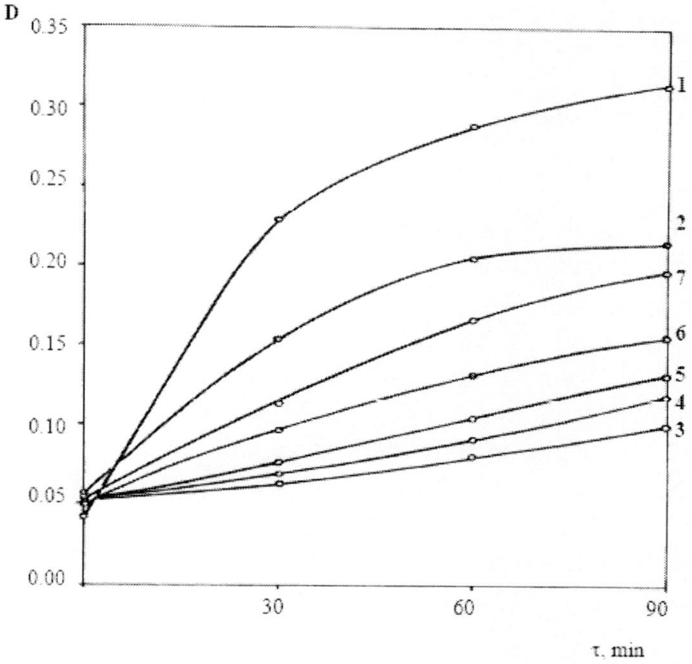

Figure 5.1. Dependence of optical density D on time of oxidation of bobbin oil AU (150°C, 1.5 hour) stabilized by composition of Ionol with *sym*-triazine XXXXc (15 : 1) at various concentrations of additive: 1 – without additive, 2 – 1 mass $ of ionol, 3 – 1 mass % of composition, 4 – 0.5 mass % of composition, 5 – 0.25 mass % of composition, 6 – 0.1 mass % of composition, 7 – 0.05 mass % of composition.

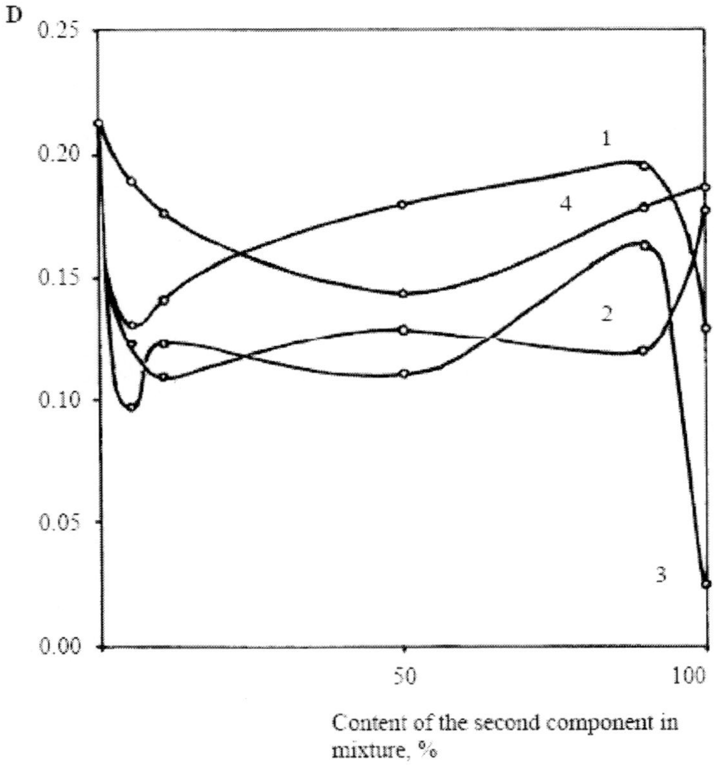

Figure 5.2. Dependence of optical density (D) of oxidized bobbin oil AU stabilized by compositional additives on the content of second components in them with Ionol (150°C, 1.5h; [InH] = 1 mass %): 1 – LXXXIIe, 2 – LXXXIp, 3 – XXXXc, 4 – XXXXd.

Optimum of components ratio in compositional additives is various in dependence on structure of studied compounds. For example for tertiary amine LXXXIIe and *sym*-triazine XXXXc the optimal ratio with Ionol is 1 :15, for derivative of piperidine LXXXIp – 1 : 10 and for *sym*-triazine XXXXd – 1 : 1. At mentioned ratios compositions reveal synergetic inhibiting effect which value in the case of compounds LXXXIp and XXXXd exceed antioxidative action of each of them. Results of carried out tests also testify to high inhibiting activity of some sulfides without Ionol exceeding it in efficiency (Table 5.6). *Sym*-triazine XXXXc reveal the highest antioxidative activity and practically completely repress oxidative processes in oil AU. Compositional additives for turbine oils used in gascompressor units were developed on the base some sulfides of *sym*-triazine. Performance of nonassociated gas fields with high content of hydrogen sulfide is connected with the problem of secure protection of gas producing and transfer equipment from corrosion and corrosion-mechanical wearing out [286].

In gascompressor units (GCU) of centrifugal type for example GCU-C-6.3/67K they usually use turbine oils TP-22 and TP-22s. Under exploitation of gascompressor units transporting sulfur-containing gas get into lubrication-jointing system of compressors gradually saturating oil by H_2S and changing its properties (acid number and sulfur content is increased, corrosion influence is strengthened). For reducing of hydrosulfide corrosion they proposed to introduce industrial inhibitors IFKhANGAZ-1, ADA, BETOL-1, AFE (0.1-0.4 mass %) into oil [287]. However these additives in the presence of corrosion-aggressive impurities in oil do not improve its antiwear and antifriction properties which may be strengthened by molybdenum-containing modifiers of friction PAF-3, TRIBOKOR-M, FRIKTOL (Mo derivatives of dithiophosphate). The last ones in their turn do not provide protecting properties of turbine oils.

Table 5.6. Change of optical density (D) under oxidation of bobbin oil stabilized by individual compounds and compositional additives with various ratios of components (150°C, [InH] = 1 mass %)

Ratio Ionol : second component in additive	Value of D before the oxidation	Value of D during the oxidation, h		
		0.5	1.0	1.5
Second component – compound LXXXIIe				
LXXXIIe	0.050	0.065	0.092	0.130
Ionol : LXXXIIe = 15 : 1	0.047	0.075	0.102	0.130
- " - 10 : 1	0.050	0.095	0.114	0.125
- " - 1 : 1	0.055	0.115	0.151	0.179
- " - 1 : 10	0.050	0.087	0.115	0.194
Second component – compound LXXXIp				
LXXXIp	0.055	0.098	0.142	0.178
Ionol : LXXXIp = 15 : 1	0.049	0.090	0.111	0.124
- " - 10 : 1	0.065	0.082	0.127	0.153
- " - 1 : 1	0.065	0.094	0.120	0.130
- " - 1 : 10	0.060	0.073	0.091	0.121
Second component – compound XXXXc				
XXXXc	0.050	0.071	0.035	0.027
Ionol : XXXXc = 15 : 1	0.055	0.062	0.083	0.098
- " - 10 : 1	0.055	0.060	0.130	0.125
- " - 1 : 1	0.050	0.058	0.094	0.112
- " - 1 : 10	0.050	0.065	0.142	0.165
Second component – compound XXXXd				
XXXXd	0.048	0.095	0.110	0.185
Ionol : XXXXd = 15 : 1	0.050	0.086	0.117	0.192
- " - 10 : 1	0.045	0.090	0.170	0.173
- " - 1 : 1	0.044	0.055	0.110	0.145
- " - 1 : 10	0.050	0.094	0.160	0.180
Ionol	0.054	0.151	0.204	0.213
Oil without additive	0.046	0.227	0.289	0.313

In this connection we studied possibility of simultaneous improvement of protecting, antiwear and antifriction properties of oil TP-22 in corrosion-aggressive medium with the help of compositions of *sym*0triazine sulfides with Mo-containing additives TRIBOKOR-M and FRIKTOL. As derivatives of *sym*-triazine we used described in Section 2.4 compounds XXXVd, XXXVe, XXXVf, XXXVi and XXXVk and also XXXVIId, e, g, h, j.

Investigation of lubricating properties of oil TP-22s was carried out on friction device MT-2M designed for estimation of oil and synthetic oils, diesel oil and special liquids in temperature interval 20-150°C [288]. We added to oil 5 vol. % of water solution of NaCl (3%) and 6g of H_2S per liter combined presence of which initiated competitive processes of formation of Fe oxides and sulfides on wear faces (at the expense of electrochemical and chemical corrosion).

We found that in structure of lubricating composition containing FRIKTOL sulfides of *sym*-triazine XXXVi, XXXVIIe and XXXVIIg reveal best properties. So, at mass ratio FRIKTOL : XXXVIIe equal to 3 : 1 deterioration is decreased by 71-73% in comparison with oil without additives. In lubricating composition FRIKTOL : XXXVIIg the lowest deterioration (by 81-83%) is reached at components ratio 1 : 3, and in composition FRICTOL + XXXVi – at ratio 1 : 1. The composition TRIBOKOR-M + XXXVIIj turned to be the most effective (deterioration is decreased by 88-91%).

Thus, introduction of compositional additives containing dithiophosphate modifiers of friction and *sym*-triazine sulfides at their optimumal ratio into oil TP-22s strengthen antiwear properties of each of the components. Simultaneously antiwear and antiwelding properties of oil TP-22s are improved: for the most effective composition mentioned above constant of friction is decreased by 92% in comparison with initial oil.

The number of *sym*-triazines was also studied as anti-oxidative additives to synthetic lubricating oil of ester type – pentaerythrite and SZhK C_5–C_9 [284]. Concentration of sulfides and known anti-oxidants (4,4'-di-*tert*-octyldiphenylamine – Octamine and phenyl-α-naphtylamine – Neozone A [289]) in oils was 1.0 mass %. Data on comparable tests are presented in Table 5.7. In accordance with data of this Table antioxidative action of the majority of tested compounds exceeds efficiency of known inhibitors Octamine and Neozone A. This fact is expressed in improvement of such exploitation indexes of ester oil as low value of acid number after oxidation, content of sediment in oxidized oil and change of kinematic viscosity. There is no corrosion on metal plates contacting with oxidized oil at all.

Table 5.7. Results of tests of thermal oxidative stability of pentaerythrite ester SZhK containing 1 mass % of *sym*-triazine sulfides and known antioxidants (Octamine and Neozone A)

Compound	Acid number after oxidation, mg KOH/g	Change (%) of kinematic viscosity of oil at temperature		Sediment after oxidation, mass %
		100°C	-40°C	
XXXVa	2.7	34.5	224	0.13
XXXVb	2.0	28.1	187	0.08
XXXVc	2.5	34.4	217	0.15
XXXVj	2.2	32.8	203	0.10
XXXVk	1.3	23.4	164	0.05
XXXVIIa	4.9	54.7	301	0.29
XXXVIIb	2.4	31.5	197	0.09
XXXVIIc	4.6	54.1	271	0.28
XXXVIIg	4.5	26.9	272	0.27
XXXVIIh	1.5	23.0	181	0.07
XXXXd	1.4	24.8	172	0.05
XXXXIVb	1.8	27.3	191	0.11
Octamine	4.1	50.3	243	0.13
Neozone A	4.9	46.1	267	0.25
Oil without additive	7.8	56.2	384	0.17

Analysis of results of tests of thermal oxidative stability of pentaerythrite ester SZhK allows disposing all studied compounds in the order of decrease of their antioxidative efficiency as follows:

XXXVk > XXXXd > XXXVIIh > XXXVb > XXXXb > XXXVIIb > XXXVj > XXXVc > XXXVa > Octamine > Neozone A > XXXVIIg > XXXVIIc > XXXVIIa

Derivatives of *sym*-triazine containing fragments of screened amine and 4-hydroxydiphenulamine – 2-alkylthio-4,6-bis(2,2,6,6-tetramethylpiperidile-4-amino)-(XXXVk), 2-alkylthio-4,6-di(4-anilinephenoxy)- (XXXXd) and 2,4-dialkylthio-6-(2,2,6,6-tetramethylpiperidile-4-amino)-*sym*-triazines (XXXVIIh) possess very high efficiency. So, under introduction of these compounds at concentration 1 mass % into testing synthetic oil the value of its acid number after high temperature (225°C, 25h) oxidation is 1.3-1.5mg of KOH/g, the amount of sediment is 0.05-0.07 mass % whereas inhibition of oil by Octamine and Neozone A increases these indexes up to 4.1-4.9mg of KOH/g and 0.13-0.25 mass % accordingly.

Sulfides of *sym*-triazine XXXVb and XXXXIVb containing in two fragments of secondary amine and 2-mercaptobenzthiazole reveal antioxidative properties strong enough. Activity of 2,4-diamino-6-alkylthio-*sym*-triazines with dimethylamino- (XXXVa), N-*n*-octadecylmethylaminogroups (XXXVc) or fragments of morfoline (XXXVj) is significantly lower, i.e. on the level of known inhibitors. We should also note that including of two higher alkyl substituents (C_{18}, C_8)into structure of *sym*-triazine derivatives positively influences on antiwear, antiwelding and anticorrosion properties of ester oil.

In this oil we also studied antioxidative efficiency of derivatives of benzimidazole, benzthiazole, 1,3,4-oxa(thia)diazole, piperidine and *sym*-triazine containing alkylthiomethyl, alkylthiomethylthiole or alkylthiomethylamine groups which synthesis and properties are described in Sections 2.3 and 4.2. Antioxidative activity of heterocyclic sulfides was estimated by the value of induction period of oxidation of ester oil with additives on manometric device, by the method of oxidation in volume on the device of bubbling type (these methods are described in experimental part) and also by change of viscosity, acid number and amount of sediment in oxidized oil and also by appearance of corrosion on metal plates.

In accordance with kinetic curves (Figure 5.3) of oxidation of ester oil with 0.02mole/l of heterocyclic sulfides at 230°C some of them posses significant inhibiting effect exceeding Octamine (induction period of oxidation = 45min). Compounds LXXXIc (Figure 5.3, curve 7, induction period of oxidation = 210min), XXIXh (curve 12, induction period of oxidation = 180min), and XXVIIf (curve 10, induction period of oxidation = 165min) possess the highest activity. Compounds XXIVb (curve 11, induction period of oxidation = 147min) and XXXe (curve 4, induction period of oxidation = 123min) reveal high antioxidative action.

Kinetic curves of O_2 absorption by ester oil at various concentration of LXXXIq in it are presented in Figure 5.4. The rise of concentration of given inhibitor up to 0.03 and 0.05mole/l (curves 5, 6) only insignificantly increases induction period of oxidation (by 15-21min); at decrease of concentration down to 0.01 and 0.005 mole/l (curves 3, 2) its induction period of oxidation is noticeably decreased (by 54-80min), whereas activity of sulfide LXXXIq is in three times higher of that one for Octamine (0.02mole/l).

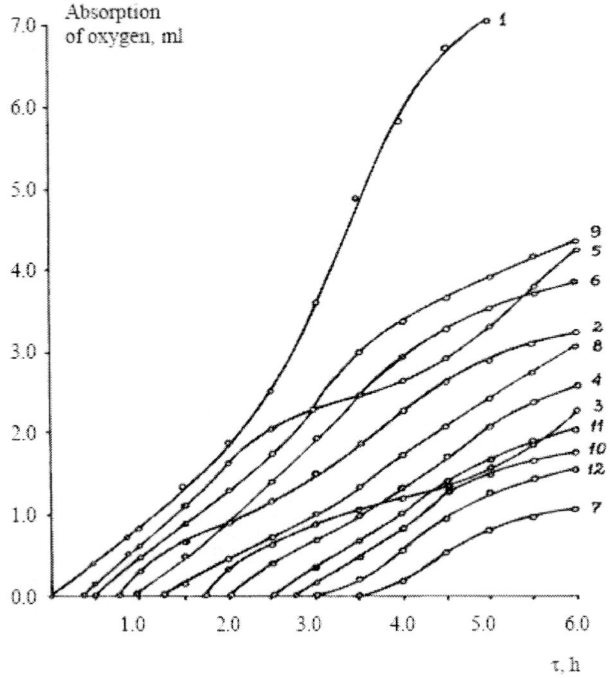

Figure 5.3. Kinetic curves of absorption of O2 at 230°C by pentaerythrite ether SZhK C5–C9 (pentaerythrite ether, 1.0g) containing antioxidative additives (0.02mole/l): 1 – pentaerythrite ether without additives, 2 – octamine, 3 – LXXXVIe; 4 – LXXXVIf; 5 – LXXXVIg; 6 – LXXXIn; 7 – LXXXIq; 8 – XIXa; 9 – XXXe; 10 – XXVIIf; 11 – XXIVb; 12 – XXIXh.

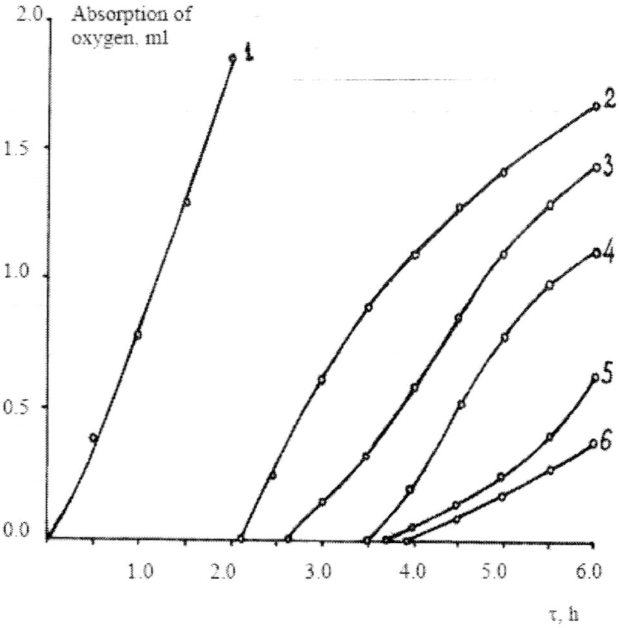

Figure 5.4. Kinetic curves of absorption of O2 at 230°C by pentaerythrite ether SZhK C5–C9 (pentaerythrite ether, 1.0g) containing sulfide LXXXIq at various concentrations, mole/l: 1 – 0; 2 – 0.005; 3 – 0.01; 4 – 0.02; 5 – 0.03; 6 – 0.05.

Under testing of samples of ester oil containing 1.0 mass %of heterocyclic sulfides their oxidation at 215°C on the device of bubbling type was carried out in the presence of aluminum (AK-4), steel (ShKh-15) and cupric (M-1) plates. Results of these tests are presented in Table 5.8.

Table 5.8. Thermal oxidative stability of pentaerythrite ether SZhK C_5–C_9 containing 1 mass %of heterocyclic sulfides (225°C, 25 hours)

Compound	Acid number after oxidation, mg/KOH	Viscosity after oxidation, mm^2/sec		Corrosion of metals, g/m^2			Sediment after oxidation, mass %
		at 100°C	at –40°C	AK-4	ShKh-15	M-1	
XIXa	2.2	6.08	24200	abs.	abs.	0.0012	0.18
XXVIIf	1.7	5.82	19210	abs.	abs.	0.0011	0.12
XXIXb	1.8	5.90	21650	0.0013	0.0007	0.0081	0.18
XXIXh	1.5	5.62	18670	abs.	abs.	0.0007	0.07
XXXe	4.8	6.72	29160	0.0020	0.0105	0.0105	0.29
LXXXIn	4.5	6.54	27680	0.0019	0.0085	0.0095	0.26
LXXXIq	1.3	5.70	18390	abs.	abs.	0.0008	0.06
LXXXVIe	1.8	5.96	22540	0.0010	0.0016	0.0018	0.11
LXXXVIf	1.9	5.97	21850	0.0009	0.0005	0.0015	0.14
LXXXVIg	4.6	6.65	28440	0.0022	0.0012	0.0095	0.27
Octamine	4.1	6.20	26700	0.0018	0.0070	0.0086	0.22
Neozone A	4.9	6.39	27800	0.0017	0.0020	0.0110	0.25
Oil without additive	7.8	7.00	38200	0.0021	0.0229	0.0110	0.15

Analysis of tests of thermal oxidative stability of ester oil containing heterocyclic sulfides allows disposing these compounds in the order of decrease of their antioxidative efficiency as follows:

LXXXIq > XXIXh > XXVIIf ≈ XXIXb > LXXXVIe > LXXXVIf > XIXa > Octamine > Neozone A > LXXXIq > LXXXVIg > XXXe

Thus, derivatives of 4-aminopiperidine LXXXIq containing fragment of screened cyclic secondary amine reveal the highest inhibiting efficiency at high temperature oxidation of ester synthetic oil. Diamino-*sym*-triazines XXIXb and XXIXh and derivatives of 1,3,4-oxadiazole XXVIIf including benzthiazolyl fragment of known stabilizer Kaptaks (or MBT) posses close to LXXXIq efficiency.

5.4. Sulfides of Sym-Triazine – Stabilizers of Hydrocarbon Fuels

Jet engine fuels obtained with the use of hydrogenation processes (hydrofining of virgin distillates and deep hydrogenation of aromatic hydrocarbons into naphthenes) posses high inclination to oxidation due to removal of heterorganic compounds being at low concentrations "natural" antioxidants out of them [290]. For stabilization of hydrogenation jet engine fuels various antioxidative additives (0.003-0.01 mass %) synthesized on the base of spatially-hindered phenols are added [291]. However efficiency of these additives due to insufficient thermal stability is sharply decreased at temperature 150°C characteristic for modern engines with higher power and calorific intensity.

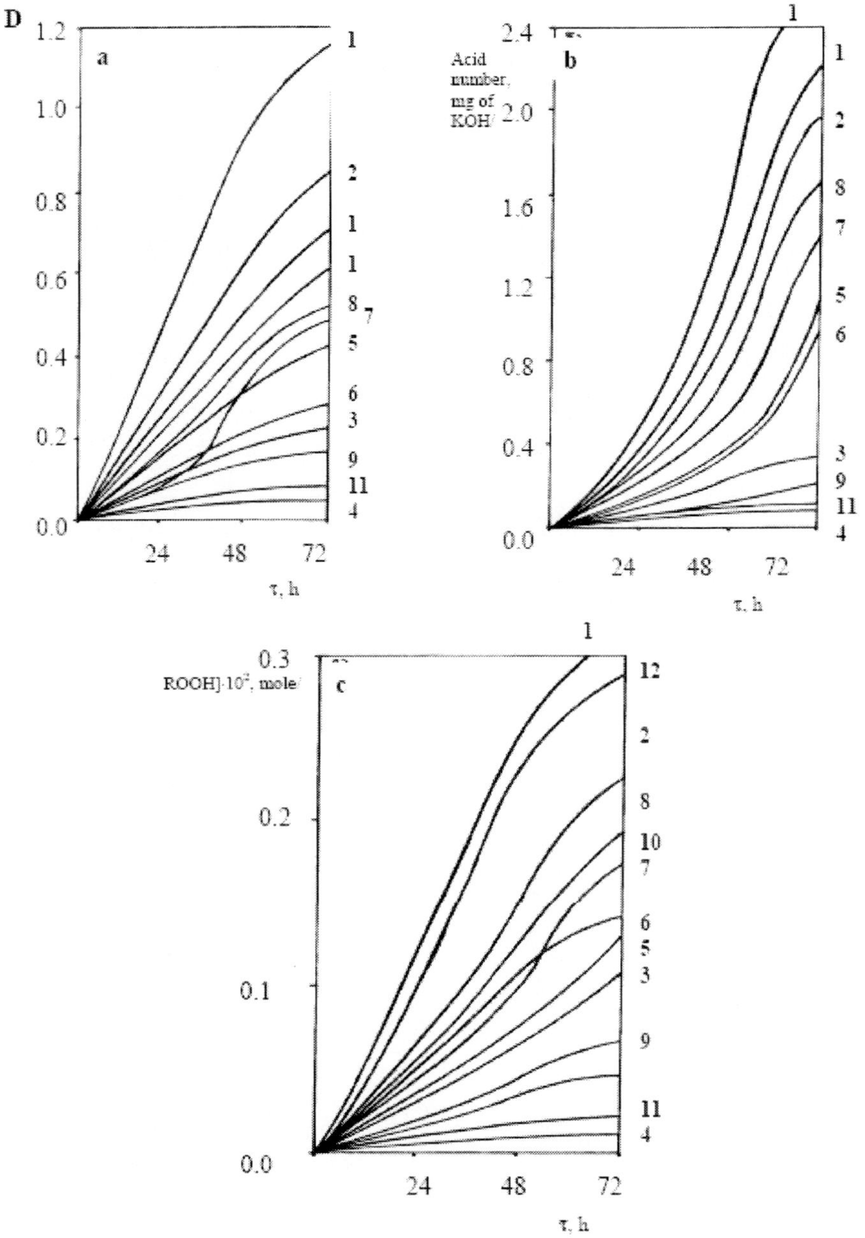

Figure 5.5. The influence of sulfides of sym-triazine (In) on optical density (D) (a), acid (b) and peroxide (c) numbers of jet engine fuels (12-repeated heating, 120°C, In = 0.003 mass %): 1 – jet engine fuel without inhibitor, 2 – Ionol, 3 – XXXVg, 4 – XXXVl, 5 – XXXVh, 6 – XXXVj, 7 – XXXVIIf, 8 – XXXVIIh, 9 – XXXXc, 10 – XXXXd, 11 – XXXXIVa, 12 – XXXXIVb.

For reception of antioxidants with higher thermal stability the derivatives of *sym*-triazine [142] including thioderivatives synthesized and tested by us as stabilizers of mineral and synthetic lubricating oils are very perspective. In presented work we studied efficiency of stabilization of jet engine fuels by sulfides of *sym*-triazine XXXVg, h, j, l, XXXVIIh, f, XXXXc, d and XXXXIVa, b.

For estimation of efficiency of tested compounds (added into jet engine fuels in amount 0.003 mass %) we used methods of 12-repeated heating of fuel at 120°C [292], its oxidability at 180°C in isolated volume [293] and measuring of sediment amount in oxidized (T = 150°C) in the presence of cupric plate fuel (on the device TSRT in accordance with GOST 11802-88). In the first case fuel was sampled every 12 hours of testing with determination of their optical density, acid and peroxide (iodometry) numbers. Antioxidative action of additives was estimated by the results of analysis (Figure 5.5) at conditions of jet engine fuels storage. The essence of the second test method which conditions are approximated to real conditions of fuels exploitation [290] is in measuring of accumulation rate and time of reaching of maximum concentration of hydroperoxides.

Derivatives of *sym*-triazine XXXVl of nonsymmetrical structure and XXXXIVa symmetrically substituted by spatially-hindered thiophenol fragments also possess antioxidative action in accordance with data of Figure 5.5. Compounds XXXXc and XXXVg, j revel lower by sufficiently high efficiency. *Sym*-triazines XXXXd and XXXXIVb are the least active.

Relatively high efficiency of compounds XXXVk, XXXVIIf, h is obviously conditioned by joint presence in their molecules of fragments of spatially-hindered phenol, substituted piperidine and sulfurorganic compounds: the first ones break radical chains of oxidation interacting with peroxide radicals, and the others destruct hydroperoxides with formation of molecular products [290]. Inhibiting effect is significantly strengthened due to the fact that phenol antioxidant reduces hydroperoxides yield under decomposition of which into free radicals oxidation chains are branched. Sulfur-containing compounds decomposing hydroperoxides prevent fast consumption of phenol inhibitor. Combination in one molecule of mentioned fragments leads to special case of synergism – to autosynergism effect [290, 291].

Results of tests of inhibiting action of sulfides of *sym*-triazine in jet engine fuels by method of oxidation in isolated volume at 180°C (Table 5.9) showed that all studied compounds increased the time of reaching of maximum concentration of hydroperoxides up to 15-58min exceeding Ionol (10min) by efficiency of action. By the influence on this time comparable antioxidants are disposed in the following order:

XXXIVa > XXXVl > XXXVg > XXXXc > XXXVh > XXXVj > XXXVIIf > XXXVIIh > XXXXd > XXXXIVb > Ionol

We found (Table 5.10) while estimating the antioxidative action of sulfides of *sym*-triazine (at concentrations 0.003 and 0.03 mass %) on the device TSRT-2 by the amount of sediment (mg/100ml) formed under oxidation of jet engine fuel for 5 hours at 150°C in the presence of cupric plate that compounds XXXXIVa, XXXVl, XXXVg, XXXXc and XXXVj possessed the highest thermal stabilizing effect (sedimentation is decreased by 82-90%). Compounds XXXXIVb and XXXXd are the least active. The interesting particularity of studied compounds is irregular decrease of sedimentation at tenfold increase of their concentration in jet engine fuels: while the efficiency of the most active additives XXXIVa and XXXVl, g is increased in 1.03-1.06 times for the least active additives (XXXXIVb, XXXXd and Ionol) it is increased in 1.33-1.40 times. Further increase of additives concentration from 0.05 and 0.1 mass % practically does not change thermal oxidative stability of jet engine fuels.

Thus, introduction into alkylthio-*sym*-triazines molecules (derivatives of natural C_2–C_5 mercaptans mixture) of two fragments of spatially-hindered phenol and screened cyclic amine allows obtaining high effective additives to jet engine fuels essentially exceeding by their characteristics industrial antioxidant Ionol.

Table 5.9. Efficiency of *sym*-triazine sulfides (In = 0.003 mass %) under oxidation of jet engine fuels in isolated volume at 180°C

Additives	The time of reaching of maximum concentration of hydroperoxides in fuel τ_{max}, min	Maximum concentration of hydroperoxides, [ROOH]·10^2, mole/l
XXXVg	52	0.21
XXXVl	56	0.18
XXXVh	40	1.04
XXXVj	35	0.87
XXXVIIf	30	1.16
XXXVIIh	24	1.42
XXXXd	21	1.54
XXXXc	49	0.25
XXXXIVb	15	1.62
XXXXIVa	58	0.14
Ionol	10	1.52

Table 5.10. The influence of *sym*-triazine sulfides on thermal oxidative stability of jet engine fules at 150°C

Additives	Sediment amount, mg/100 ml	Decrease of sedimentation, %
XXXVg	4.5/3.1*)	83.7/88.8
XXXVl	4.0/2.7	85.6/90.3
XXXVh	9.5/7.5	66.1/72.9
XXXVj	7.8/5.6	71.9/79.8
XXXVIIf	7.4/4.2	73.3/84.8
XXXVIIh	13.7/8.6	50.2/68.7
XXXXc	5.1/3.4	81.6/87.7
XXXXIVb	14.8/10.5	46.6/62.1
XXXXd	16.0/11.4	42.2/58.9
XXXXIVa	3.6/2.8	87.0/89.9
Ionol	13.5/8.8	50.9/68.0
Fuel without additives before oxidation	27.7	-

Note: *) In numerator – concentration of additive is 0.003 mass %, in denominator – 0.03 mass %

Finally we studied the efficiency of a number of sulfides of *sym*-triazine in composition of additives for stabilization of compounded diesel fuels. Such fuels prepared by mixing of virgin oil distillates with light gas oil of catalytic cracking and similar products of oil secondary refining are widely used in Russia together with diesel fuels of weighted and broadened fraction structure for broadening of fuel resources [291, 294, 295].

However compounded diesel fuels are characterized by higher inclination to oxidation, resinification and sedimentation [296, 297] at conditions of storage and exploitation. Their stabilization by traditional inhibitors (Ionol, OMI, Agidol-12, FCh-16, PODFA, etc.) is insufficient that is explained by significant amount of impurities of unsaturated and polyaromatic hydrocarbons and also heteratom compounds revealing pro-oxidative action [295-297]. Effective stabilization of mix diesel fuels may be realized by compositional additives containing antioxidants, metals deactivators, dispersants and tertiary aliphatic amines (the last ones neutralize acid impurities and primary products of hydrocarbons oxidation) [291, 297].

In given work we studied possibility of stabilization of mix diesel fuel by sulfides of *sym*-triazines XXXVb, g, h, j, XXXVIIb, f, g, XXXXc, d, XXXIVa, b and also by their compositions with N,N'-dimethyl-4-hydroxy-3,5-di-*tert*-butylbenzylamine (antioxidant OMI). It is an industrial antioxidant [289], it is highly effective under stabilization of light gas oil of catalytic cracking and its mixtures with virgin diesel fuels [291, 295, 297]; sulfides of *sym*-triazine XXXVg, h, XXXVIIf, XXXXc, d, XXXIVa, b reveal significant antioxidation action in jet engine fuels as it was shown above in this Section.

The mixture of virgin diesel fuel of brand L (GOST 305-82) with 20% of light gas oil of catalytic cracking (fraction with boiling temperature 195-301°C from the plant G-43-107 of Moscow petroleum refinery) was used as subject of inquiry. For estimation of stabilizers efficiency we studied the kinetics of accumulation of hydroperoxides in this diesel fuel with additives (0.01 mass %). The fuel was oxidized by air oxygen at 140°C for 150 min in reactor of bubble type and the content of hydroperoxides was measured every 15 minutes [290] by the method of iodometric titration (Figure 5.6).

As it is obvious from the data of Figure 5.6 diesel fuels without additives (curve 1) and with 0.01 mass % of OMI are oxidized intensively enough (even with OMI additive oxidized fuel contains $8.4{\cdot}10^{-4}$ moles of hydroperoxides/l). Some of tested sulfides of *sym*-triazines in comparison with OMI significantly more active in inhibition of oxidation in diesel fuels. The observed oscillatory character of kinetic curves is obviously connected with accumulation of hydroperoxides, their decomposition and formation of products being corrosion inhibitors.

Compounds XXXXc (curve 7), XXXVg (curve 4), XXXXIVa (curve 9) and also XXXVIIf (curve 12) including fragments of spatially-hindered phenol possess the highest efficiency. For these sulfides concentration of accumulated hydroperoxides does not exceed $(0.66\text{-}0.42){\cdot}10^{-4}$mole/l. Compounds XXXVh (curve 5) and XXXXd (curve 8) also reveal inhibiting effect strong enough, whereas sulfides XXXVj, XXXXIVb, XXXVIIb and XXXVIIg (curves 6, 10, 11, 13) are inferior to OMI.

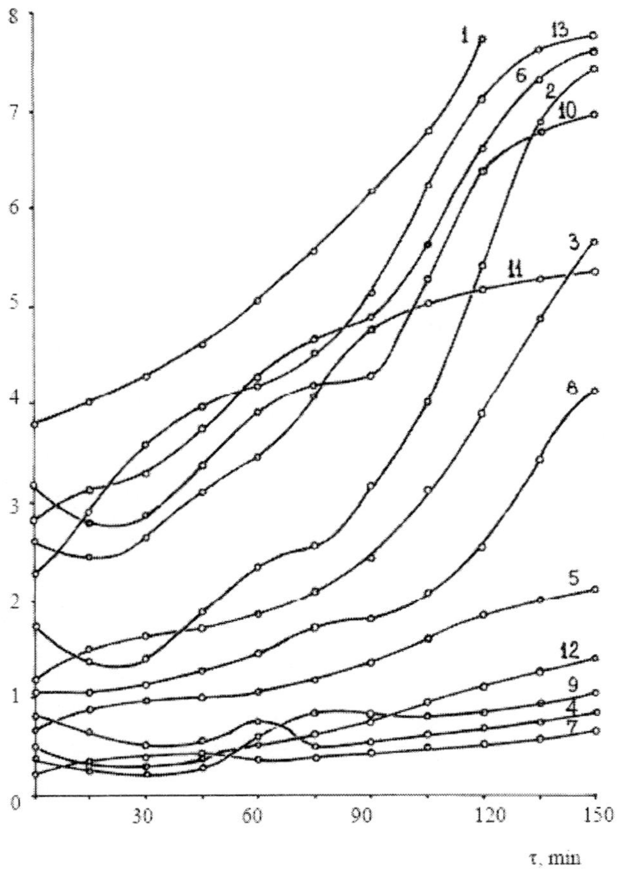

Figure 5.6. The influence of sym-triazine sulfides (0.01%) on kinetics of accumulation of hydroperoxides under oxidation of compounded diesel fuel: 1 – fuel without additives; 2 – OMI; 3 – XXXVb; 4 – XXXVg; 5 – XXXVh; 6 – XXXVj; 7 - XXXXc; 8 – XXXXd; 9 – XXXXIVa; 10 – XXXXIVb; 11 – XXXVIIb; 12 – XXXVIIf; 13 – XXXVIIg.

Tested compounds by the efficiency in reducing of hydroperoxide concentrations may de disposed in the following order:

XXXXc > XXXVg > XXXXIVa > XXXVIIf > XXXVh > XXXXd > XXXVb > OMI > XXXVIIb > XXXXIVb > XXXVj > XXXVIIg

Then we studied antioxidative efficiency of listed sulfides in their compositions with OMI at mass ratio 1 : 10 at total concentration in diesel fuel 0.01 mass%. Results of carried out tests presented in Figure 5.7 showed that these compositions decreased concentration of hydroperoxides in the following consequence:

OMI + XXXVg > OMI + XXXXIVa > OMI + XXXVh > OMI + XXXXc > OMI + XXXXVId > OMI + XXXVIIf > > OMI + XXXXIVb > OMI + XXXVb > OMI + XXXVIIb > OMI > OMI + XXXVj > OMI + XXXVIIg

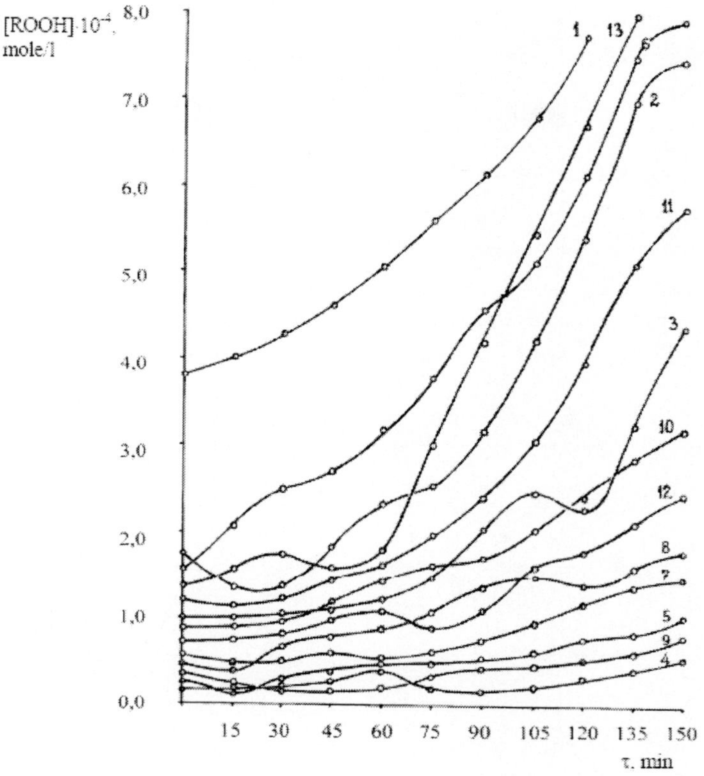

Figure 5.7. Influence of antioxidant OMI compositions (0.01 mass %) with derivatives of sym-triazine (10 : 1) on kinetics of accumulation of hydroperoxides under oxidation of compounded diesel fuel: 1 – fuel without additive; 2 – fuel with OMI additive; 3 – OMI + XXVb; 4 – OMI + XXXVg; 5 – OMI + XXXVh; 6 – OMI + XXXVj; 7 – OMI + XXXXc; 8 – OMI + XXXXd; 9 – OMI + XXXXIVa; 10 – OMI + XXXXIVb; 11 – OMI + XXXVIIb; 12 – OMI + XXXVIIf; 13 – OMI + XXXVIIg.

For estimation of influence of compositional additives of this raw on stability of compounded diesel fuel we also determined acid number, content of insoluble sediment and actual resins in accordance with GOST 11802-86 (Table 5.11). Fuel was oxidized by air oxygen for 16 hours at 100°C in the presence of plate of electrolytic copper as it was specified in the complex of methods of qualification analysis [293]. In accordance with obtained data the amounts of formed after oxidation sediment and actual resins and also acids in fuel with compositional additives in the majority of cases are lower than in fuel without additives or containing only OMI.

Compositions of OMI with *sym*-triazine sulfides including two fragments of spatially-hindered phenol are the most effective (compounds XXXVg, XXXXc and XXXXIVa). They practically completely prevent sedimentation, decrase the content of actual resins in oxidized fuel in 2.1-3.3 times and acid content in 3.1-4.7 times (Table 5.11). Compositions of OMI with XXXVh and XXXXd also possess high inhibiting action.

Efficiency of stabilization of mix diesel fuel by the most perspective compositional additives containing sulfides XXXVg and XXXXIVa was estimated in dependence on their composition and total concentration (Table 5.12).

Table 5.11. The influence of compositional additives on thermal oxidative stability of compounded diesel fuel (100°C, 16 hours, additives content = 0.01mass %, OMI : *sym*-triazines = 10 : 1)

Additives	Sediment, mg/100 mlл	Actual resins, mg/100 ml		Acid number, mg KOH/100 ml	
		in initial diesel fuel	in diesel fuel after oxidation	in initial diesel fuel	in diesel fuel after oxidation
OMI	4.88	35	65	0.28	1.30
OMI +XXXVb	2.88	30	58	0.22	0.84
OMI +XXXVg	отс.	18	32	0.12	0.37
OMI +XXXVh	отс.	21	35	0.16	0.48
OMI +XXXVj	4.96	40	69	0.30	1.25
OMI +XXXXc	abs.	24	38	0.15	0.56
OMI +XXXXd	1.78	26	32	0.21	0.39
OMI +XXXXIVa	abs.	15	24	0.10	0.41
OMI +XXXXIVb	2.50	27	48	0.19	0.65
OMI +XXXVIIb	4.80	42	67	0.32	1.18
OMI +XXXVIIf	2.25	25	43	0.27	0.44
OMI +XXXVIIg	4.77	39	68	0.31	1.34
diesel fuel without additives	7.53	41	79	0.35	1.75

Table 5.12. The influence of concentration and components ratio of compositional additives on thermal oxidative stability of compounded diesel fuel (100°C, 16 hours)

Ratio OMI : *sym*-triazines in additive (mass %)	Sediment mg/100 ml	Actual resins, mg/100 ml		Acid number, mg KOH/100 ml	
		in initial diesel fuel	in diesel fuel after oxidation	in initial diesel fuel	in diesel fuel after oxidation
OMI, 0.01	4.88	35	65	0.28	1.30
OMI, 0.05	3.93	31	56	0.21	1.21
OMI + XXXVg (15:1), 0.01	1.75	24	41	0.18	0.64
OMI + XXXVg (5:1), 0.01	abs.	16	30	0.14	0.31
OMI + XXXVg (10:1), 0.05	abs.	13	23	0.13	0.25
OMI + XXXVg (15:1), 0.05	1.18	22	46	0.22	0.72
OMI + XXXVg (5:1), 0.05	abs.	15	29	0.15	0.33
OMI + XXXXIVa (15:1), 0.01	1.04	20	35	0.17	0.51
OMI + XXXXIVa (5:1), 0.01	abs.	16	21	0.10	0.30
OMI + XXXXIVa (10:1), 0.05	abs.	18	30	0.16	0.54
OMI + XXXXIVa (15:1), 0.05	0.94	21	34	0.18	0.42
OMI + XXXXIVa (5:1), 0.05	abs.	12	21	0.22	0.50
Diesel fuel without aditives	7.53	41	79	0.35	1.75

Change of OMI-to-compound XXXVg (or XXXXIVa) ratio from 10 : 1 up to 15 : 1 increases content of sediment and actual resins, and change from 10 : 1 down to 5 : 1 only insignificantly decreases these parameters. With the rise of additive concentration from 0.01 up to 0.05mass % (components ratio is 10 : 1) in the case of XXXXIVa content of actual resins and acid number of oxidized fuel are essentially increased; when we used XXXVg stabilization effect was strengthened even more (Table 5.12).

Thus for more effective stabilization of mixture of virgin diesel fuel with 20% of light gas oil of catalytic cracking it is advisable to introduce into it composition (10 : 1) of industrial antioxidant OMI with sulfides of *sym*-triazine including fragments of screened phenol or 4-aminodiphenylamine.

5.5. Biological Activity of Substances Synthesized on the Base of C_2–C_5 Mercaptans

It is known [7, 49, 58, 60, 174, 191, 192, 195, 222, 227, 230, 235, 258, 268] that a lot of heterocyclic compounds possess high and versatile biological activity including derivatives of imidazole, benzazole, oxadiazole, thiadiazole, *sym*-triazine, 1,3-dioxalane, etc. which synthesis and properties are intensively studied at present [226, 298-304]. In accordance with investigations carried out in I.M. Gubkin Russian State University [305-308] introduction of sulfur atoms into heterocycles and a number of aliphatic compounds in the majority of cases broadens spectrum of their biological activity and increases it.

In connection with this fact tests of antimicrobial, chemotherapeutical and pesticide action of various types of compounds synthesized by us on the base of C_2–C_5 mercaptans are of great scientific and practical interest. Such tests were carried out in laboratories of All-Russian scientific centre on safety of drugs (Kupavna city), Ufa scientific-research technological institute of herbicides and regulators of plants growth, V.A. Nikitin Institute of chemistry of Tadjikistan Academy of Sciences.

5.5.1. Biocide Additives to Hydrocarbon Oils

Antiseptic additives – biocides are used for fight with biological affection of hydrocarbon substrates including lubricating oils [309, 310]. Various heterorganic compounds in particular Sn-organic derivatives of alkylphenol [311], S-, O-derivatives of spatially-hindered alkylphenol [312] and some others [309, 310] are used as biocide additives as biocide additives.

Results of investigations carried out in I.M. Gubkin Russian State University [307, 308] showed that individual sulfur-containing derivatives of N-acylhydrazones, thiazolidine-4-ones and chloropropanols possessed effective antimicrobial action in relation to widespread in oil products mold fungus and bacteria (*Aspergillus niger, Candida tropicalis, Pseudomonas aczudinosa*). That is why we carried out analogous tests for compounds synthesized on the base of SMK namely XIIIc, d, LId, h, i, j and also LXXVIc, d soluble in base oil of selective purification M-8 [284].

As the estimation method we used "lune" method on agaric medium [311] with the use of suspension of mentioned above pure cultures of mold fungus and bacteria affecting oil products and aggressively destructing them [309, 310]. Degree of biocide effect of studied compounds was estimated by the value of diameter of zone of fungus and bacteria suppression around "lune" with additive (0.25-1.0 mass %): the higher this value the higher the efficiency of additive. The oil M-8 does not possess biostability.

According to results of tests (Table 5.13) among studied samples compounds XIIIc, d reveal the highest antibacterial and antifungal activity, compounds LXXVIc, d and LIi are less active, efficiency of compounds LIh, j, d is small. It is also obvious that by biocide action compounds XIIIc, d are on one level with earlier studied 4-hydroxy-3,5-di-*tert*-butylbenzaldehyde N-indolyl-3-acetylhydrazone [307] and 4-hydroxy-3,5-di-*tert*-butylbenzaldehyde benzthiazolylthioacetylhydrazone [308] revealed ones of the most highest antimicrobial and fungicide properties.

Table 5.13. Antimicrobial activity of compounds in oil M-18

Compounds	Concentration in oil, mass %	Width of zone of suppression of growth of microorganisms around the "lune", cm*)		
		Pseudomonas aczudinosa	*Candida tropicalis*	*Aspergillus niger*
XIIIc N-(β-C_2-C_5-alkylthioacetyl)hydrazones of 4-HO-3,5-di-*tert*-C_4-benzaldehude	1.0 0.5 0.25	1.3-1.5 1.2-1.3 1.0-1.0	2.6-2.8 2.1-2.2 1.2-1.2	4.5-4.8 3.3-3.5 2.3-2.5
XIIId N-(β-C_2-C_5-alkylthioacetyl)hydrazones of 5-nitrofuryl aldehyde	1.0 0.5 0.25	1.2-1.3 1.0-1.1 1.0-1.0	2.5-2.7 2.1-2.2 1.0-1.2	4.3-4.7 3.2-3.4 2.2-2.4
LId 1-(C_2-C_5-alkylthio)-3-*n*-octadecylmethylaminopropane-2	1.0 0.5 0.25	1.0-1.0 1.0-1.0 1.0-1.0	1.4-1.5 1.1-1.2 1.0-1.0	2.6-2.7 2.0-2.1 1.5-1.6
LIh 1-(C_2-C_5-alkylthio)-3-(4-HO-3,5-di-*tert*-C_4-anilino)propanols-2	1.0 0.5 0.25	1.0-1.1 1.0-1.0 1.0-1.0	1.8-1.9 1.4-1.5 1.0-1.1	2.8-2.9 2.4-2.5 1.8-2.0
LIj 1-(C_2-C_5-alkylthio)-3-piperidino-propanols-2	1.0 0.5 0.25	1.2-1.3 1.1-1.2 1.0-1.0	1.9-2.1 1.6-1.7 1.1-1.3	3.0-3.2 2.6-2.7 2.1-2.3
LIj 1-(C_2-C_5-alkylthio)-3-morfolino-propanols-2	1.0 0.5 0.25	1.0-1.0 1.0-1.0 1.0-1.0	1.9-2.0 1.3-1.5 1.0-1.0	2.7-2.8 2.1-2.2 1.6-1.7
LXXVIc 2-(4-HO-3,5-di-*tert*.-C_4-phenyl-2-H-3-[N(β-C_2-C_5-alkylthiopropionyl)amino]thiazolidine-4-ones	1.0 0.5 0.25	1.4-1.6 1.2-1.3 1.0-1.0	2.7-2.9 2.0-2.1 1.2-1.2	3.7-3.8 3.0-3.1 2.4-2.5
LXXVId 2-(5-nitrofuryl-2)-2-H-[N(β-C_2-C_5-alkylthiopropionyl)amino]thiazolidines-4-ones	1.0 0.5 0.25	1.2-1.3 1.0-1.1 1.0-1.0	2.5-2.7 2.2-2.3 1.8-1.9	3.5-3.6 2.7-2.8 2.3-2.4
N-Indolyl-3-acetylhydrazones of 4-HO-3,5-di-*tert*-C_4-benzaldehyde [307]	1.0 0.5 0.25	1.4-1.5 1.0-1.0 1.0-1.0	2.7-2.9 2.0-2.1 1.2-1.2	4.7-4.9 3.5-3.7 2.4-2.6
Benzthiazolylthioacetylhydrazone of 4-HO-3,5-di-*tert*-C_4-benzaldehyde [308]	1.0 0.5 0.25	1.3-1.4 1.1-1.2 1.0-1.1	2.8-3.0 2.2-2.4 1.2-1.2	4.6-4.8 3.6-3.7 2.5-2.7

Note: *) in accordance with standard technique recommended by GOST [313].

Thus, compounds XIIIc, d, LXXVIc, d and LIi may serve as potential high effective additives to mineral oils, they are able to provide their stability to bio-failures revealing antibacterial and especially fungicide activity. Growth of fungus even the most aggressive ones (*Aspergillus niger*) destructing oil products is completely suppressed by mentioned compounds synthesized on the base of C_2–C_5 mercaptans mixture.

5.5.2. Inhibition of Biological Degradation of Surface-Active Substances Applied for Intensification of Oil Production by Microorganisms of Oil-Field Waters

As it is known [314] at modern technology of oil (and also hydrocarbon condensate) production in productive stratums there large amounts of mineralized water containing large quantity of chlorides (up to 250g/l and more) and sulfides of sodium, calcium, magnesium, iron, etc., and also anaerobic microorganisms including sulfate-reducing bacteria [315]. The last ones transform inorganic sulfates into H_2S causing strong corrosion effect on metal of oil producing equipment.

Moreover sulfate-reducing bacteria are able to subject to biodegradation a lot of chemical reagents used for intensification of production of oil, natural gas and gas condensate [316, 317], in particular surfactant species, for example often applied sodium alkylarylsulfonates and alkylphenoxypolyethoxylates. That is why protection of chemical reagents from biodestruction with the aim of maintaining of their technological effect is necessary.

At present in domestic industrial practice of oil production they use sodium salt of isopropylphenol [318], quaternary ammonium bases such as alkyltrimethylammonium chloride, alkylpolybenzylpyridinium chloride [132], and bases of more complex structure – trietyhl-6-(benzo-1,4-dioxanomethyl)ammoniumchloride [319], tri-*n*-butyl-6-(2-methylbenzo-1,4-dioxanomethyl)ammoniumchloride [320], etc. as reagents for suppression of biodestruction.

In our work we studied biocide activity of a number of synthesized heterocyclic compounds of various types in relation to sulfate-reducing bacteria. They are: 2-R-2-R^1-3-[N-(β-C_2-C_5-alkylthiopropionyl)amino]thiazolidine-4-ones LXXVIc, d, 3-(β-C_2-C_5-alkylthiopropionyl)-4-R-1H-1,2,4-triazoline-5-thiones LXXVIIb and 2-R-2-(2-β-C_2-C_5-alkylthio-1-R^1-ethyl)-4-R^2-1,3-dioxolanes LXXXXVe, g. Test were carried out in accordance with known technique.

Under uncontaminated conditions we introduce 1 mass % of surface active agent Sulfanol NP-1 (sodium alkylarylsulfonate) or Neonol AF 9-12 (nonylphenoxypolyethoxylate) and also 1-2,5 mass % of tested compound into oil-field water of Orenburg gas condensate field containing 236g/l of mineral salts and 10^5–10^6cells/ml of sulfate-reducing bacteria. Then we take samples (1 ml) and introduce them into bottles with 10 ml of Postgeit nutrient medium, thermostate them at 32°C for 10, 15, 20 and 25 days. By the end of mentioned times in samples (and also in control sample) the surface tension (σ) is determined by the method of count of solutions drops on stalagmometer [320] and by the content of formed H_2S by conductometric titration by 5% water solution of NaOH. The values of decrease of σ and H_2S content of tested samples (in comparison with control one) presented in Table 5.14 characterize activity of biocides presenting in them.

It is obvious from the obtained results that all tested biocides possess high degree of suppression of sulfate-reducing bacteria vital functions in stratum water: formation of H_2S is decelerated by 70-100% for 25 days. For the first 10 days of tests almost all biocides except LXXVId and LXXVIIb at concentration 0.1 mass % (or 0.25 mass %) completely protect surface active agent from biodestruction.

Table 5.14. Results of tests of biocide activity of heterocyclic derivatives of mercaptans C_2–C_5 mixture

Biocide; concentration in solution, mass %	Surface active agent, mass %	Time of tests, days	σ, mN/m	Degree of biodestruction of surface active agent, %	Degree of suppression of bacterial formation of H_2S, %
LXXVIc; 0.1	Sulfonol NP-1, 0.1	10	35.2	0	100
		15	36.9	4.8	95
		20	38.7	9.9	90
		25	40.5	15.0	83
	Neonol AF 9-12, 0.1	10	32.5	0	100
		15	33.6	3.4	97
		20	34.5	6.2	92
		25	35.8	10.2	88
LXXVIc; 0.25	Sulfonol NP-1, 0.1	15	35.2	0	100
		20	36.1	3.1	100
		25	37.4	6.3	100
	Neonol AF 9-12, 0.1	15	32.5	0	100
		20	33.3	2.5	100
		25	33.8	4.0	100
LXXVId; 0.1	Sulfonol NP-1, 0.1	10	35.3	0.3	100
		15	37.2	5.7	95
		20	39.1	11.1	90
		25	40.9	16.2	82
	Neonol AF 9-12, 0.1	10	32.5	0	100
		15	33.8	4.0	95
		20	34.9	7.4	90
		25	36.2	11.4	85
LXXVId; 0.25	Sulfonol NP-1, 0.1	15	35.2	0	100
		20	36.6	4.0	100
		25	37.8	7.4	100
	Neonol AF 9-12, 0.1	15	32.5	0	100
		20	33.6	3.4	100
		25	34.5	6.2	100
LXXVIIb; 0.1	Sulfonol NP-1, 0.1	10	35.5	0.9	98
		15	37.9	7.7	95
		20	40.2	14.2	88
		25	42.3	20.2	80
	Neonol AF 9-12, 0.1	15	34.2	5.2	93
		20	35.3	8.6	88
		25	36.8	13.2	82
LXXVIIb; 0.25	Sulfonol NP-1, 0.1	15	35.6	1.1	100
		20	37.1	5.4	90
		25	38.3	8.8	80
	Neonol AF 9-12, 0.1	15	33.4	2.8	100
		20	34.7	6.8	95
		25	36.1	11.1	85
LXXXXVe; 0.1	Sulfonol NP-1, 0.1	10	35.2	0	100
		15	35.8	1.7	98
		20	37.2	5.1	95
		25	38.4	9.1	90
	Neonol AF 9-12, 0.1	10	32.5	0	100
		15	32.9	1.2	98
		20	34.2	5.2	95
		25	35.3	8.6	90

Table 5.14. Continued

Biocide; concentration in solution, mass %	Surface active agent, mass %	Time of tests, days	σ, mN/m	Degree of biodestruction of surface active agent, %	Degree of suppression of bacterial formation of H_2S, %
LXXXXVe; 0.25	Sulfonol NP-1, 0.1	15	35.2	0	100
		20	35.3	0.3	100
		25	37.3	6.0	100
	Neonol AF 9-12, 0.1	15	32.5	0	100
		20	32.5	0	100
		25	33.8	4.0	100
LXXXXVg; 0.1	Sulfonol NP-1, 0.1	10	35.2	0	100
		15	35.6	1.1	98
		20	36.9	4.8	95
		25	38.0	8.0	90
	Neonol AF 9-12, 0.1	10	32.5	0	100
		15	32.8	0.9	98
		20	34.0	4.6	95
		25	35.2	8.3	90
LXXXXVg; 0.25	Sulfonol NP-1, 0.1	15	35.2	0	100
		20	35.2	0	100
		25	37.2	5.7	100
	Neonol AF 9-12, 0.1	15	32.5	0	100
		20	32.5	0	100
		25	33.5	3.1	100
Katapin-bactericide; 0.25	Sulfonol NP-1, 0.1	10	35.2	0	100
		15	40.5	15.0	85
		20	44.1	25.3	70
	Neonol AF 9-12, 0.1	10	32.5	0	100
		15	36.2	11.4	85
		20	38.8	19.4	72
Cetylpyridinium chloride; 0.25	Sulfonol NP-1, 0,1	10	35.2	0	100
		15	40.1	13.9	90
		20	41.5	17.9	83
	Neonol AF 9-12, 0.1	10	32.5	0	100
		15	35.4	8.9	90
		20	37.6	15.7	85

At this concentration biodestruction of surface active agent reveal only after 15 days, in the case of most active biocides LXXXXVe, g – only by 1.1-1.7%, after 20 days – by 4.6-5.7% and only after 25 days – by 8.0-9.1%; for the least active biocides LXVIIb, LXXVId, c already after 20 days 9.9-14.2% of Sulfanol NP-1 are destructed, Neonol AF 9-12 for the same period is destructed for 6.2-8.6% (in other words Neonol AF 9-12 is more stable to biodestruction than Sulfanol NP-1).

At concentration 0.25 mass % of the most active biocides LXXXXVe, g, LXXVIc, d in tested samples biodestruction of surface active agent for 10 days is insignificant – for Sulfanol NP-1 it is 0–4%, for Neonol AF 9-12 – 0-3.4%; for check standard – cetylpyridinium chloride and Katapin-bactericide – biodestruction of surface active agent for 20 days occur to a greater extent – by 15.7-25.3%.

Thus, synthesized by us heterocyclic derivatives SMK protect surface active agents from biodestruction better than known biocides. The highest activity in this case is revealed by compounds LXXXXVg, e, compounds LXXVIc, d and LXXVIIb are less active in comparison with cetylpyridinium chloride and Katapin-biocide.

5.5.3. Chemotherapeutical Action of S, N, O-Derivatives of C_2–C_5 Mercaptans Mixture

One of the main problems of medical science is still the fight with infected microbes and fungus (staphylococcus, streptococcus, blue pus bacillus, colon bacillus, colibacillus, Escherichia coli and soporiferous bacillus) causing numerous illnesses [321]. In this fight various preparations are used, mainly antimicrobe ones hindering growth and propagation of microbes or destroying them and also neutralizing toxic products of their vital functions. The most effective preparations are antibiotics however microbes with time become stable to their action. That is why the search of more biologically active preparations is constantly in the field of vision of researches working in given field.

Compounds chosen for tests (substituted alkylthioethanols VIIg, h, alkylthiopropanols LIi, j, N-hydrazones of aldehydes LXVIc, d, thiazolidine-4-ones LXXVIc, d, f, thiosemicarbazones LXXXXc, d, f, g, i) were preliminary characterized by the index of acute toxicity (LD_{50}). It was studied in the experiment on white mousse of both sexes with mass 20-25g at single introduction of substances in composition of 3-5% water suspension – 1ml [322]. The values of LD_{50} were determined by modified method with the use of statistical treatment [323]. Practically all studied compounds turned to be low toxic (LD_{50} = 810-1480mg/kg).

Antibacterial and antifungal activities were determined *in vitro* by degree of suppression of growth of staphylococcus, streptococcus, colon bacillus, salmonella and number of other fungus. The tests were carried out by the method of twofold serial cultivation on liquid nutrient medium: Khottinger both (for bacteria) and Saburo medium (for fungus). Microbe load in test with bacteria was 1×10^5 KOE/ml, with fungus – 1×10^6 KOE/ml. Bacteria were incubated at 37°C for 18 hours, fungus – at 25°C for 24 hours. Compounds were tested at concentration ≤ 250 mcg/ml.

The criterion of bactericide activity was minimum limit concentration of tested compound at which there was no bacteria growth for 5 days in comparison with control, and fungicide activity was estimated by compound concentration at which there was no fungus growth for 30 days. Results of tests were compared (Table 5.15) with the action of widely used with this aim ethanol, phenol, chloroamine, nitrofungine and Rivanol [322].

In accordance with data presented in Table 5.15 almost all tested compounds possess antimicrobe and antifungal activities. Compounds VIIg, h, LIj, LXVIc, LXXVIf, LXXXXc, d, f, g, i should be assigned to moderately acting preparations, LIi, LXVId, LXXVIc, d to strongly functioning which represent practical interest from the point of view of their application in veterinary and medicine.

For some of tested compounds antiinflammatory activity on white rats with mass 200-240g was studied with introduction of 0.1 ml of 1% water solution of carragenine into their back foot. Growth of inflamed foot was estimated oncometrically before and after 1 hour after introduction of phlogogenic agent. Orthophene and lysine acetylsalicylate were used as standard for comparing. All studied compounds and standards (doze is 10mcg/kg) were introduced intraperitoneally as 15 water suspension 1 hour before injection of carragenine. The data on decrease of inflammation edema (in comparison with standard) measured every 1-4 hours (averaging values) and values calculated on the base of experimental data of ED_{50} – doze causing 50% hindering of foot edema are presented in Table 5.16.

Table 5.15. Antibacterial and antifungal activity of S, N, O-derivatives of C_2–C_5 mercaptans mixture

No.	LD_{50}	Minimum limit concentration of compounds *in vitro*							
		Microbes					Fungus		
		Staphylococ-cus aureas	Bac. Subtilis	Escherichia Coli	Pr. Vulgaris	Ps.Aeruginosa	Microsporium Canis	Trichphyton Gypseum	C. Albicans
VIIg	870	125	> 250	125	250	125	250	125	250
VIIh	1250	125	> 250	125	125	125	> 250	250	125
LIi	1050	62.5	250	125	62.5	62.5	125	250	125
LIj	930	125	250	125	125	125	125	125	125
LXVIc	900	125	125	> 250	> 250	> 250	> 250	> 250	> 250
LXVId	850	31	62.5	> 250	> 250	250	31	62.5	62.5
LXXVIc	1480	31	31	62.5	> 250	> 250	62.5	62.5	62.5
LXXVId	845	15.5	15.5	62.5	125	> 250	31	62.5	62.5
LXXVIf	1000	125	125	> 250	> 250	> 250	62.5	62.5	31
LXXXXc	1430	125	125	> 250	> 250	> 250	> 250	250	> 250
LXXXXd	1250	125	125	> 250	> 250	> 250	> 250	> 250	250
LXXXXf	915	125	125	125	> 250	> 250	125	125	> 250
LXXXXg	930	125	125	125	> 250	> 250	125	125	> 250
LXXXXi	845	125	> 250	> 250	> 250	> 250	125	62.5	31
Standard preparations									
Ethanol		> 250							
Phenol (1%)		> 250							250
Chloroamine (1%)		> 250							
Nitrofungine (1%)		> 250							250
Rivanol, solution (1:1000)		> 250					125	125	250

Table 5.16. Antiinflammatory activity of some S, N, O-derivatives of C_2–C_5 mercaptans

Compound	LD_{50}, mg/kg	Hindering of edema (% to standard)	ED_{50}, mg/kg
LXVIc	900	35.15	123.6
LXVId	850	23.56	145.8
LXXIIb	840	40.04	81.4
LXXIId	1170	68.54	33.5
LXXIIe	810	50.35	60.4
LXXVIIf	890	63.44	41.4
LXXXXc	1430	12.64	-
LXXXXf	915	16.05	-
LXXXXg	930	8.58	-
CIVc	1080	20.35	-
CIVe	910	15.84	-

It is obvious that compounds LXXXXc, f, g and CIVc, e practically do not possess antiinflammatory activity or reveal weak action. Somewhat higher activity is observed for compounds LXVIc, d and LXXIIb, e, and maximum for compounds LXXIId. f.

Anticonvulsant activity of a number of heterocyclic compounds was studied on models of cramps caused by pentylenetetrazol and strychnine on white mouses. studied compounds were introduced intraperitoneally in increasing dozes as water suspension.

Results of tests (Table 5.17) showed that at therapeutic doses of *sym*-triazines XXIXb most effectively change the character of convulsive stroke (tonic phase of convulsions disappears that completely prevent animals death), derivatives of benzimidazoles XXIXb, a and 1,3,4-oxadiazole XXVIIb, f functioned to a less degree. Compounds XIXa, XXVIIb and XXIXb reveal the same activity against strychnine effect.

Antitumoral and antileukemic actions of a number of heterocycles LXXIId, e, f, LXXVIIa-c and aliphatic S, N, O-containing compounds XIIIc, d were also determined.

Investigations were carried out on white nonlinear he-rats with mass 120-130g which were inoculated with Yensen sarcoma and mousse of BDF_1 line with mass 20-25g which were infected by lymphoid leucaemia L-1210; we used *o*-methylpharagonine as standard for comparison [324]. Preparations were introduced intraperitoneally once a day for 8 days, anticancer activity was estimated by indexes of tumor hindering ($J_T = (B_c - B_e)/B_c$, where B_c and B_e – the average masses of tumor in control and experimental groups of animals) in experiments of Yensen sarcom. Antileukemic activity was estimated by the increase of life time of mousse of experimental group in comparison with control one.

Among investigated compounds (Table 5.18) the highest antitumoral effect is revealed by compounds LXXIId, f and LXXIXc for which J_T is 58-60% at maximum doze. The same compounds reveal high antileukemic effect: the rise of animals lifetime reaches 45-47% (for standard only 32%). The rest tested compounds (XIIIc, d, LXXVIIa, b) are some what inferior to mentioned preparations.

Table 5.17. Anticonvulsant activity of heterocyclic compounds

Compounds		LD_{50}, mg/kg	ED_{50}, mg/kg	
No.	Name		Anticorrozole activity	Antistrychnine activity
XIXa	2-(β-C_2-C_5-alkylthiomethyl)-benzimidazole	1055	204.5	292.0
XIXb	2-(β-C_2-C_5-alkylthiomethyl)-5-chlorobenzimidazole	1215	174.0	154.5
XXVIIb	2-(β-C_2-C_5-alkylthiomethyl)-5-nitro-1,3,4-oxadiazole	1415	207.0	188.0
XXVIIf	2-(β-C_2-C_5-alkylthiomethyl)-5-(indolyl-3)-1,3,4-oxadiazole	1134	216.0	280.5
XXIXb	2-dimethyamino-4-amino-6-(β-C_2-C_5-alkylthiomethyl)-*sym*-triazine	1335	64.5	192.6

We also studied radiation protective efficiency of thiazolidine-4-ones LXXVIc, d, N,N-disubstituted C_2–C_5-alkylthiomethylamines LXXXIk, m, n, N-substituted di(C_2–C_5-alkylthiomethylamines) LXXXIIe, g, h and substituted thiosemicarbazones LXXXXf, i. Tests were carried out on mouses hybrids (CBA × $C57B_1)F_1$ with mass 19-23g, into abdominal cavity of which we introduced water suspension (0.2-0.4ml) of tested compounds in 30 minutes before or enterally in 1 hour before the irradiation. The known radioprotector 2-aminoethylisothiouronium served as comparison object [325]. Animals were irradiated by isotope ^{137}Cs on IGUR device at dose of 850 r-units and power 210 r-units/min (at one and same time).

Radioprotection activity of compounds was estimated by probability of survival of animals for 30 days after irradiation. Death of control animals under given conditions was 100%. As it obvious from data of Table 5.19 the majority of tested compounds reveal expressed or good radioprotective activity (25-69% at 50-300mg/kg). The best radioprotectors (72-81%) under intraperitoneal introduction are compounds LXXVIc and LXXXXi, and under enteral one are LXXXIk and LXXXIIe; compounds LXXVId, LXXXIn, LXXXIIh are less active; compounds LXXXIm, LXXXIIg, h, LXXXXf possess relatively small activity.

For tests on mutagenic activity we choose compounds revealing the highest anti-inflammatory and antitumoral action LXXIId, f, LXXIXc, and also radioprotective action – LXXVIc, LXXXXd. Antimutagenic activity was studied by the method of "doze-effect" on biochemical cultures *Escherichia coli* and *Actimyces rimosus* [326]. Activity of compounds was estimated by frequency of revertants popularity from auxotrophic to prototrophic states by locuses responsible for synthesis of threonine and lysine. Mutations appearing spontaneously serve as control [327].

The influence of chosen compounds on mutations induced by UV-irradiation (bactericide lamp BUV-30) was studied on the same test-objects. Experiments were carried out in darkened box at red light. Control was made by frequency of mutations appearing under treatment (for 10 minutes) of objects containing known protectors – 2-mercaptoethylamine and cystamine (doze was 0.1mg). Results of experiments are presented in Table 5.20.

Studying of genetic activity of tested compounds showed that it does not depend on both structure and used test-object. Tested compounds and control protectors reveal mutagenic activity: in threonine locus of colon bacillus induction of mutation is lower in comparison with spontaneous one in control experiements; in lysine locus of antimicetes compounds LXXIId practically do not possess protection activity, and compounds LXXVIc and LXXIXc reveal activity high enough (on the level of standards).

Under mutations induced by UV-irradiation compounds LXXVIc and LXXIXc to a greater extent decrease their development in the case of test-cultures decreasing the number of mutation by 70-79% (for *E. coli* P-678) and by 38-44% (for *Act. rimosus* 222). The rest tested compounds either reveal equal to positive control action, or are inferior to it.

Thus, carried out tests of a number of synthesized by us S, N, O-derivatives of C_2–C_5-mercaptans mixture allow revealing among them perspective compounds by their physiological activity (XXVIIb, g, XXIXb, LIi, j, LXXIId, f, LXXVIc, d, LXXIXc, LXXXXi) which action (in accordance with laboratory investigations) exceed control preparations. These compounds are recommended for further broaden tests.

Table 5.18. Antitumor and antileukemic activities of some heterocyclic compounds

Compounds		LD_{50}, mg/kg	Doze, mg/kg	Yensen sarcoma, J_T, %	Lymphatic leukemia (lifetime, %)
No.	Name				
XIIIc	N-(β-C_2-C_5-alkylthioacetyl)hydrazones of 4-HO-3,5-di-*tert*-C_4-benzaldehyde	1320	30 25	40 25	40 27
XIIId	N-(β-C_2-C_5-alkylthioacetyl)hydrazones of 5-nitrofuryl aldehyde	1030	40 25	50 30	40 22
LXXIId	2-(β-C_2-C_5-alkylthioethyl)-4-(4-HO-3,5-di-*tert*-C_4-phenylmethylene)-Δ^2-imidazoline-5-thiones	840	30 25 15	60 40 25	47 42 35
LXXIIf	2-(β-C_2-C_5-alkylthioethyl)-4-(pyridile-3-methylene)-Δ^2-imidazoline-5-thiones	1170	30 25 15	58 40 22	45 38 30
LXXIIa	3-(β-C_2-C_5-alkylthioethyl)-4-propenyl-1H-1,2,4-triazole-5-thiones	850	30 25	35 23	32 24
LXXIXb	3-(β-C_2-C_5-alkylthioethyl)-4-propenyl-5-benzeneamine-4H-1,2,4-triazoles	970	40 25	40 28	40 25
LXXIXc	3-(β-C_2-C_5-alkylthioethyl)-5-acetoxy-4H-1,2,4-triazoles	1410	40 25	58 40	45 40
o-methylfagaronine (standard for comparison)		550	25	50	32

Table 5.19. Radioprotective activity of S, N, O-derivatives of C_2–C_5 mercaptans mixture

Compounds		LD_{50}, mg/kg	Way of introduction*)	Doze, mg/kg	Exceeding of animals lifetimes (over control), %
No.	Name				
LXXVIc	2-(4-HO-3,5-di-*tert*-C_4-phenyl)-2-H-3-[N-(β-C_2-C_5-alkylthiopropionyl)amino]thiazolidine-4-ones	1480	A B	150 250	61 49
LXXVId	2-(5-nitrofuryl-2)-2-H-3-[N-(β-C_2-C_5-alkylthiopropionyl)amino]-thiazolidine-4-ones	845	A B	200 250	63 52
LXXXIk	N-(4-HO-3,5-di-*tert*-C_4-phenyl)-N-(β-C_2-C_5-alkylthiomethyl)amines	940	A B	100 200	57 65
LXXXIm	N-(pyridile-2)-N-(β-C_2-C_5-alkylthiomethyl)amines	825	A B	200 250	25 40
LXXXIn	N-(thiazolyl-2)-N-(β-C_2-C_5-alkylthiomethyl)amines	815	A B	200 200	38 32
LXXXIIe	N-di(β-C_2-C_5-alkylthiomethyl)-N-(4-HO-3,5-di-*tert*-C_4-phenyl)-amines	1120	A B	200 300	58 69
LXXXIIg	N-di(β-C_2-C_5-alkylthiomethyl)-N-(pyridile-2)amines	960	A B	200 200	35 40
LXXXIIh	N-(β-C_2-C_5-alkylthiomethyl)-N-(thiazolyl-2)amines	835	A B	200 200	53 37
LXXXXf	1-(β-C_2-C_5-alkylthio)-3-phenylpropylthiosemicarbazones	915	A B	150 250	31 26
LXXXXi	1-(β-C_2-C_5-alkylthio)-3-(furyl-2)-propyl-4-phenylthiosemicarbazones	1250	A B	150 250	69 55
2-aminoethylthiouronium (standard for comparison)		480	A B	400 400	36 48

Note: *) A – intraperitoneal introduction, B – enteral introduction.

Table 5.20. The influence of S, N, O-derivatives of mercaptans on probability of survival of cells and appearance of mutations of test-objects

Compounds		LD$_{50}$, мг/кг	*Escherichia coli* P-678			*Actinomuces rimosus* 222		
No.	Name		Probability of survival, %	Population of revertants/10^6 survived cells		Probability of survival, %	Population of revertants/10^6 survived cells	
				absolute	% to control		absolute	% to control
Natural mutation (control)			100	8.9±0.70	100	100	5.0±0.45	100
LXXIId	2-(β-C_2-C_5-alkylthioethyl)-4-(4-HO-3,5-di-*tert*-C_4-phenylmethylene)-Δ^2-imidazoline-5-ones	1170	92	8.3±0.45	96	78	5.9±0.60	118
LXXIIf	2-(β-C_2-C_5-alkylthiomethyl)-4-(pyridile-3-methylene)-Δ^2-imidazoline-5-ones	890	87	7.2±0.80	84	113	4.0±0.30	80
LXXVIc	2-(4-HO-3,5-di-*tert*-C_4-phenyl)-2-H-3-[N-(β-C_2-C_5-alkylthiopropionyl)amino]thiazolidine-4-ones	1480	108	4.8±0.50	56	105	3.2±0.45	64
LXXIXc	3-(β-C_2-C_5-alkylthioethyl)-4-propenyl-5-acetoxy-4H-1,2,4-triazoles	1410	106	5.1±0.80	59	110	2.9±0.45	58
LXXXXi	1-(β-C_2-C_5-alkylthio)-3-(furyl-2)propyl-4-phenylthiosemicarbazones	845	90	6.4±0.45	74	102	3.6±0.30	72
2-mercaptoethylamine			90	5.4±0.45	62	55	2.7±0.30	54
Cystamine			97	8.4±0.45	98	101	2.8±0.30	55
UV-mutation (control)			100	6.0±0.50	100	100	5.0±0.40	100
LXXIId	2-(β-C_2-C_5-alkylthioethyl)-4-(4-HO-3,5-di-*tert*-C_4-phenylmethylene)-Δ^2-imidazoline-5-ones	1170	124	5.2±0.20	87	57	7.8±0.90	155
LXXIIf	2-(β-C_2-C_5-alkylthiomethyl)-4-(pyridile-3-methylene)-Δ^2-imidazoline-5-ones	890	60	4.5±0.35	75	107	5.8±0.80	116
LXXVIc	2-(4-HO-3,5-di-*tert*-C_4-phenyl)-2-H-3-[N-(β-C_2-C_5-alkylthiopropionyl)amino]thiazolidine-4-ones	1480	134	1.5±0.10	25	65	3.0±0.35	60
LXXIXc	3-(β-C_2-C_5-alkylthioethyl)-4-propenyl-5-acetoxy-4H-1,2,4-triazoles	1410	150	1.8±0.50	30	148	3.1±0.20	62
LXXXXi	1-(β-C_2-C_5-alkylthio)-3-(furyl-2)propyl-4-phenylthiosemicarbazones	845	76	3.9±0.45	65	94	3.8±0.30	76
2-mercaptoethylamine			172	1,4±0,04	22.5	168	3.0±0.40	60.2
Cystamine			83	6,5±0,80	108.5	122	3.5±0.50	70.4

Table 5.21. Fungicidal activity of some S, N, O-derivatives of C_2–C_5 mercaptans mixture

Compounds		LD_{50}, mg/kg	Test-objects (% of suppression of diseases)				
No.	Name		Aspergillus Penicillium	Sclerospora graminicola	Phytophtorose of tomato	Farinaceous dew of cucumbers	Gray mould of beans
VIa	1-(C_2-C_5-alkylthio)-2-methoxymethoxyethanes	1365	0	44	38	68	32
VIIg	1-(C_2-C_5-alkylthio)-2-morfolinomethoxyethanes	870	68	27	98	93	54
XXa	2-(β-C_2-C_5-alkylthiomethyl)-benzoxazoles	1130	28	73	94.5	81	92
XXb	2-(β-C_2-C_5-alkylthiomethyl-6-nitro)- benzoxazoles	900	40	52	31	18	8.5
XXVIIb	2-(β-C_2-C_5-alkylthiomethyl)-5-nitro-1,3,4-oxadiazoles	1450	0	10	52	40	62
XXVIIf	2-(β-C_2-C_5-alkylthiomethyl)-5-(thiophene-2)-1,3,4-oxadiazoles	1134	0	0	82	53	11.5
Standards: Tiram [48, p.280]		400	100	100	100	-	-
Euparen [48, p.255]		~2000	-	-	-	-	100
Dinokap [48, p.252]		~700	-	-	-	100	-

Table 5.22. Results of tests of fungicide activity of some S, N, O-derivatives of C_2–C_5-mercaptans mixture in relation to fungus disease of cotton

Compound		LD_{50}, mg/kg	Concentration by active substance, mass %	Plants diseases, %		
No.	Name			wilt	root rot	marcescence
VIIg	1-(C_2-C_5-alkylthio)-2-morfolinomethoxyethanes	870	0.1 0.01 0.001	0 2.8 7.5	0 9.9 22.0	0 11.5 23.3
XXb	2-(β-C_2-C_5-alkylthiomethyl)-5-nitro-6-H-benzoxazoles	900	0.1 0.01 0.001	0 1.7 6.4	0 8.6 14.4	0 10.3 19.0
XXIXb	2-amino-4-dimethylamino-6-(β-C_2-C_5-alkylthiomethyl)-*sym*-triazines	1335	0.1 0.01 0.001	0 1.1 5.0	0 7.4 13.1	0 9.5 17.2
XXXXc	2,4-bis(4-HO-3,5-di-*tert*-C_4-phenyloxy)-6-C_2-C_5-alkylthio-*sym*-triazine	945	0.1 0.01 0.001	0 0 3.6	0 5.3 11.2	0 4.8 12.3
XXXXIVa	2,4-*bis*(4-HO-3,5-di-*tert*-C_4-phenylthio)-6-C_2-C_5-alkylthio-*sym*-triazine	1140	0.1 0.01 0.001	0 0 3.0	0 2.9 5.1	0 3.1 7.9
XXXXIVb	2,4-*bis*(benzthiazolyl-2-thio)-6-C_2-C_5-alkylthio-*sym*-triazine	1033	0.1 0.01 0.001	0 1.2 5.8	0 4.9 14.1	0 0 1.7
Эталон: Benomil [48, p.239]		~8000	0,1 0,01 0,001	0 0 33	0 3.3 28	0 4.5 24
Control (without pesticides)			-	82.5	78	91

5.5.4. Pesticide Activity of S, N, O-Derivatives of C_2–C_5 Mercaptans Mixture

Tests of a number of synthesized by us S, N, O-derivatives of C_2–C_5 mercaptans mixture by specialists of scientific-research technological institute of herbicides and regulators of plants growth (Ufa) and department of plants protection of Tajik agricultural college showed that compounds Via, VIIg, XXa, b and XXVIIb, g possess essential fungicidal effect on exciters of farinaceous dew of cucumbers and some other plants diseases (Table 5.21). Compounds 1-(C_2-C_5-alkylthio)-2-morfolinomethoxyethanes VIIg and 2-(β-C_2-C_5-alkylthiomethyl)benzoxazoles XXa reveal the highest activity in relation to tomato phytophtorose exciters approaching the activity of standard (Table 5.21).

When estimating the counteraction to fungus disease of cotton-plant we established (Table 5.22) that all tested compounds at concentration 0.1 mass % (in calculation per active substance) to 100% suppressed mentioned diseases; compounds XXXXIVa, XXXXc and XXIXb at concentrations 0.001-0.01 mass % suppress infection by wilt by 95-100%, black root rot bu 87-95%, the rest compounds in their activity are close to standard (Table 5.22).

We found that among 20 tested by herbicide activity under laboratory conditions on etiolated germs of bioindicators (oats, garden radish, millet, cucumbers) compounds XXIXb, XXXXIVa, b reveal the highest phytotoxicity (more than 75%).

The number of compounds was tested by growth-regulating activity under laboratory conditions on etiolated wheat, cucumbers and tomato germs. The most effective were compounds XIXa, XXVIIg and also XXXVg, i: they accelerated growth and development of underground and overground parts of vegetable plants. Some preparations were checked at field conditions under tomato sowing (Table 5.23). They showed that *sym*-triazines XXXVg, I by 27-40% (in comparison with control) increase their early crop capacity to a greater extent than known growth-stimulators – gibberelline and heteroauxin.

Consequently as a result of tests of pesticide activity of various S, N, O-derivatives of C_2–C_5 mercaptans mixture the compounds were revealed possessing high fungicidal, herbicidal and growth-regulating activity in relation to widespread vegetable and cereal plants.

Table 5.23. The influence of some substituted *sym*-triazines on total and early harvest of tomato under field conditions

Compounds		LD_{50}, mg/kg	Harvest from one shrub	
No.	Name		early	total
XXXVg	2,4-*bis*(4-HO-3,5-di-*tert*-C_4-phenylamino)-6-C_2-C_5-alkylthio-*sym*-triazine	1195	0.75/128.4*)	21.2/2079.3
XXXVi	2,4-*bis*(thiazolil-2-amino)-6-C_2-C_5-alkylthio-*sym*-triazine	886	0.72/126.5	20.5/1896.8
Heteroauxin [48, p.36]		~300	0.85/121.0	19.0/1836.8
Gibberelline [48, c. 36]		6300	0.62/76.8	18.2/1637.5
Control (water without growth stimulators)			0.44/72.8	22.1/1932.3

Note: *) quantity of fruitages/ fruitages mass, g (average statistical data)

Thus, carried out investigations of applied properties of a great number of organic compounds which were synthesized by us on the base of C_2–C_5 mercaptans mixture showed that among them there were compounds able to be applied as effective polyfunctional additives to mineral oils and synthetic lubricating oils, stabilizers of hydrocarbon fuels and also biologically active substances with varied antimicrobe, chemotherapeutical and pesticide activity in relation to many test-objects.

APPENDIX: METHODS OF CARRYING OUT EXPERIMENTS

Fraction of C_2–C_5 mercaptans with melting temperature 35-130°C isolated from odorant of Orenburg gas refinery (yield 83-86% in calculation per odorant) was used as initial raw material. In accordance with gas-liquid chromatography this fraction contains, vol. %: *n*-hexane – 2.0; *n*-heptane – 0.5; benzene – 1.0; toluene – 1.0; athanthiole – 4.0; *n*-propanethiole-1 – 20.5; *n*-propanthiole-2 – 38.5; 2-methylpropanthiole-2 – 1.0, *n*-butanthiole-2 – 22.0; *n*-butanthiole-1 – 5.0; 3-methylbutanthiole-1 – 3.5 and *n*-pentanethiole-1 – 1.0. The average molecular mass of fraction 84.6 (it was determined by ebullioscopic method); $\mathbf{n_D^{20}}$ 1.4421; $\mathbf{d_{20}^{4}}$ 0.840 g/cm^3.

The separating funnel of volume 250ml is filled by 100ml of orenburg gas condensate with density 0.712g/cm^3 including 0.020 mass % of H_2S and 1.941 mass % of mercaptans C_1–C_6 and 30ml of water solution of diethanolamine (25 mass %), they are stirred at temperature 20°C for 5 minutes and settle. The lower (water) layer containing dissolved salt H_2S and DEA is isolated, and to the 99ml of upper (hydrocarbon) layer 50ml of water solution of sodium hydroxide (10 mass %) containing 0.04-3.86g of di-*n*-butylphthalate or 0.01-0.95g pyrazole and 0.05-19.3g *n*-propanol is added for the rise of additives solubility.

After extraction, settling (for 10 min) and further separation of layers purified from H_2S and mercaptans the gas condensate and also condensate of mercaptans (1.287-1.366g) acidulous up to pH = 7 by diluted hydrochloric acid is dried by calcined $CaCl_2$ and analyzed by gas-liquid chromatography on Orenburg gas refinery [146] of the device LKhM-8 MD in column 4000x3mm filled by bearer celite C-545+20% reoplex 400 at temperature 90°C and consumption of gas-bearer helium 3.6l/h.

In accordance with results of gas-liquid chromatography analysis of initial and demercaptanized gas condensate, isolated from it mixture of C_1–C_6 mercaptans their residual content $\mathbf{C_k}$ and degree of isolation $\boldsymbol{\eta}$ are calculated (as ratio of amount of isolated thioles to initial amount) which values were presented in Table 2.1 and 2.2.

Solution of *n*-propan- and *n*-hexanethiols and also of dibutylphthalate and pyrazole in methanol with equal molar concentration (for *n*-propanthiole 1.043mole/l, for *n*-hexanethiole 0.4955mole/l) are prepared for carrying out of spectral investigations. Index of refraction of these solutions is determined on refractometer [328] at 20°C (n_D^{20}) with accuracy up to fifth sign, and then curves of $(\mathbf{n_D^{20}})^2$ dependences on thioles-to-activating additives ratio in solution are plotted (Figure 2.1 – for dibutylphthalate; Figure 2.3 – for pyrazole).

For the same solutions PMR spectra were obtained on device "Tesla-80" with the use of HMDS as internal standard. In spectra of these solutions the shift of signal of thiole group SH into weak field with the rise of dibutylphthalate and pyrazole concentrations occurs (Table 2.3 and 2.4) which is changed disproportionately testifying to formation of molecular associates.

Detailed component structure of a number of isolated of Orenburg gas-refinery group of marketable mercaptans and products of their distillation is determined with the help of chromatography-mass spectrometry on device "Finnigan MAT-1125" in glass capillary column at the energy of ionized irradiation 80eV.

Distillate fractions of mercaptans (odorants) were tested on intensity and stability of smell according to GOST 22387.5-77 by organoleptic method in odorizer in accordance with 5 points scale. For estimation of odorant inclination to oxidation the following technique was developed. Micronized powder of Fe_3O_4, water and odorant were placed into glass vessel, air was bubbled through liquid layer at room temperature for 4 hours. By the end of tests contents of vessel was divided into layers and the upper layer (odorant) was analyzed with the help of gas-liquid chromatography on thioles and dialkyldisulfides contents, the lower (water) layer was analyzed by the content of dissolved in it thioles. Results of analysis and estimations of inclination of odorant samples to oxidation are presented in Table 2.8.

Anti-corrosion properties of mineral oil M-11 containing S-alkyl ethers of thiocarbonic acids or N- and P-derivatives of 1-alkylthiopropanol-2 were determined by mass loss of leaden plates (g/m^2) on the device DK-NAMI at 140°C for 25 hours by the technique reported in GOST 20502-75. Antiwear and antiwelding properties of these samples of oil were estimated accordingly by the diameter of wear scar ($\mathbf{D_w}$) and tearing index ($\mathbf{I_t}$) on four-ball tester of friction of *Shell* firm by GOST 9490-75 with balls of diameter 9.52mm of steel ShKh-15at constant load on upper ball 200H (rotational frequency of spindle was 1420min^{-1}, relative rate of slipping of friction surfaces was 0.54m/sec, time of testing was 60 min).

Thermal oxidative stability of oil with additives was determined by K.K. Papok method at 250°C (GOST 23175-78). Evaluative parameter was time (min) during which thin layer of oil was transformed into film.

Anticorrosion (protective) properties of mineral oils I-40 and M-11 containing sulfides of *sym*-triazine were determined by GOST 9.054-75, methods 1, 4 and 5. Tests by method 1 were carried out in hygrothermal camera G-4 in regime of periodic condensation of moisture on samples of steel St 20. At first samples were subjected to the influence of air at temperature 40±2°C and relative humidity 95±3% for 7 hours. Then conditions of moisture on samples were created by cooling of them down to temperature lower than temperature of camera by 5-10°C. In the course of testing samples were visually tested every equal time intervals from the beginning of tests until appearance of the first corrosion center.

Tests by method 4 were carried out at constant immersion of steel sample into electrolyte (pH=8), by method 5 – under the effect of 0.1% solution of HBr acid. Additives were introduced into oils at concentrations 0.5 and 1 mass % and compared by efficiency with known inhibitors of corrosion Penazolin 10-16 [132].

Antioxidative properties of synthetic ester oil (pentaerythric ether SZhK C_5–C_9) were studied by method of oxidation in volume on the device of bubble type by GOST 2379-79 in the presence of aluminum (AK-4), steel (ShKh-15) and cupric (C-1) plates under the current of dry air with the rate 170ml/min for 15 hours at 225°C. Acid number after oxidation of oil (mg of KOH/g), change of kinematic viscosity of the last one and amount of sediment were estimated by techniques reported in [284, 293].

Moreover, efficiency of antioxidative action of heterocyclic sulfides in synthetic ester oil was studied on manometric device with its evaluation by the value of induction period of oxidation [281, 282]. The method consists in oxidation of oil samples by oxygen under statistical conditions at 230°C for 6 hours. Known inhibitors of oxidation Octamine and Neozone A were used as objects for comparison [289].

For estimation of efficiency of stabilizers of hydrocarbon fuels we used standard methods: of 12-fold heating of fuel at 120°C [292], its oxidability at 180°C in isolated volume [293] and measuring of sediment amount in oxidized (**T** = 150°C) in the presence of cupric plate fuel (on the device TSRT by GOST 11802-88), i.e. techniques reported in the collection of qualification test methods [293]. For estimation of oxidability of diesel fuels we studied kinetics of accumulation of hydroperoxides in it by technique reported in the book [290]: fuel containing additives (0.01 mass %) were oxidized by air oxygen at 140°C for 150min in reactor of bubble type and every 15 min we determined the content of hydroperoxides by known method of iodometric titration.

For determination o efficiency of biocide action of additives in hydrocarbon oils we used standard technique desribed in GOST 9.048-75 – 9.052.75.

Under investigation of chemotherapeutical action of synthesized compounds we estimated acute toxicity ($\mathbf{LD_{50}}$) by known method [7], antibacterial and antifungal activity and also other parameters of pharmacological effect were estimated by techniques described in special literature [323, 325, 326] and applied in medical practice.

Pesticide activity of tested compounds was determined in accordance with recommendations and methods reported in monograph [7] and reference book [48].

REFERENCES

[1] Karrer P. Course of organic chemistry, transl. from German, ed. by Kolosov M.N. – Leningrad: *Goskhimizdat,* 1960, p.153-155.

[2] Yanovich A.N., Asvatzaturov A.Tz., Busurin A.A. Protection of labour and safety measures in gas equipment – Moscow: *Nedra,* 1978, 316p.

[3] Gritzenko A.I., Akopova G.S., Maksimov V.M. Ecology. Oil and gas. – Moscow: *Nauka*, 1997, 597p.

[4] Kan R., Dermer O. Introduction into chemical nomenclature, transl. from English, ed. by Potapov V.M., Lidin R.A. – Moscow: *Khimiya*, 1983, 148p.

[5] Korshunov M.A., Mazaev V.E. Processes of reception and application of mercaptans in industry (subject review) – Moscow: *TzNIITEneftekhim*, 1967, 30p.

[6] Grunvald V.R., Afanasiev Yu.M. Preparation and processing of gas and gas condensate (information review) – Moscow: *VNIIGasprom*, 1986, No.8, 43p.

[7] Melnikov N.N. Pesticides. Chemistry, technology and application – Moscow: *Khimiya*, 1987, 712p.

[8] Foigt I. Stabilization of synthetic polymers against light and heat effect, transl. from German, ed. by Kovarskaya B.M. – Leningrad: *Khimiya*, 1972, p.272-288.

[9] Oae S. Chemistry of sulfur organic compounds – transl. from Japan, ed. by Prilezhaeva E.N. – Moscow: *Khimiya,* 1975, 512p.

[10] Barret J.K. Thioles, in book by Barton D., Wollis U.D. General organic chemistry, transl. from English, ed. by Kochetkova N.K., Nifantieva E.E. – Moscow: *Khimiya*, 1985, V.5, p.130-187.

[11] Krasnov K.S. Molecules and chemical bond – Moscow: *Vysshaya shkola*, 1977, 280p.

[12] Chemistry of sulfur organic compounds, ed. by Belenkii L.I., Bzhezovskii V.M., Vlasova N.N., et. al. – Moscow: *Khimiya,* 1988, p.9-22.

[13] Bernardi F., Csizmadia I.G., Mangini A. Organic Sulfur Chemistry. Theoretical and Experimental Advances. – Amsterdam: *Elsevier*, 1985, 740p.

[14] Zuika I.V., Bankovskii Yu.A. *Uspekhi khimii*, 1973, 42, No.1, p.39-64.

[15] Lyapina N.K. Chemistry and physico-chemistry of sulfur-organic compounds of oil distillates – Moscow: *Nauka*, 1984, 120p.

[16] Lyapina N.K., Lygin V.I., Ulendeeva A.D., *Uspekhi khimii*, 1980, 49, No.9, p.998-1013.

[17] Bolshakov G.F. Sulfur-containing compounds of oil – Novosibirsk: *Nauka*, 1986, 246p.

[18] Chertkov Ya.B., Spirkin V.G. Sulfur and oxygen compounds of oil distillates – Moscow: *Khimiya,* 1971, 307p.

[19] Shenfeld N., Surface active agents on the base pf ethylene oxide, transl. from German, ed. by Lebedev N.N. – Moscow: *Khimiya*, 1982, p.60-61.

[20] Kortland C, Bortslap C. *Tenside*, 1968, 5, No.2, s.43-44.

[21] Böhme H., Fischer H., Frank R. *Preparation and properties of α-halogenated thio ethers –Ann*, 1949, 563, s.54-72.

[22] Jirousek L., Kostic J.V. *Chem. Listy*, 1949, 43, No.2, p.105-109.

[23] Broniarz J., Szymanovski J., Pernac J., Pujanek M., *Przem. chem.*, 1977, 56, No.1, p.17-22.

[24] Turchaninova L.P., Korchevin N.A., Shipov A.G. et. al. *Zh. ob. khimii*, 1989, 59, No.3, p. 722-723.

[25] Prochukhan A.S., Baranovskaya E.M., Lukmanova A.S. *Organic compounds of sulfur. Collection of articles* – Riga: Zinatne, 1980, 2, p.204-206.

[26] Lewin S. *J. Chem. Soc.*, 1962, p.1462-1467.

[27] Levi T.Y. *Gaz. Chim. ital.*, 1932, 62, p.775-780.

[28] Lichtenthaler F.W., Voss P. *Justus Liebigs Ann. Chem.*, 1969, 724, s.81-90.

[29] Poppelsdorf F., Holt S.J. *J. Chem. Soc.*, 1954, p.1124-1130.

[30] Pollack I.E., Grillot G.F. *J. Org. Chem.*, 1967, 32, No.10, p.2892-2895.

[31] Dronov V.I., Nigmatullina R.F., Krivonogov V.P. *J. Org. Chem.*, 1976, 12, No.8, p.1672-1674.

[32] Patent 51-25004 Japan 1976.

[33] Burckhalter J.H., Wells J.H., Mayer W.J. *Tetrahedron Letters.*, 1964, No.21, p.1353-1359.

[34] Böhme H., Heller P. *Chem Ber.*, 1953, 86, s.443-450.

[35] Gill N.S., James K.B., Liones F., Potts K.T. *J. Amer. Chem. Soc.*, 1952, 74, No.19, p.1871-1874.

[36] Prilezhaeva E.N., Shostakovskii M.F. *Uspekhi khimii*, 1963, No.8, p.897-947.

[37] Lyapina N.K., Tolstikov G.A., Melnikova L.A. *Neftekhimiya*, 1983, 23, No.3, p.403-408.

[38] Prilezhaeva E.N., Pivovarova T.E., V,I, Snegotzkii, et. al. Chemistry of sulfur organic compounds containing of oil and oil products: Collection of articles – Moscow: *Vysshaya shkola,* 1972, 9, p.198-207.

[39] Heiba E.-A.I. *J. Org. Chem.*, 1966, 31, No.3, p.776-780.

[40] Dmuchovsky B., Zienty F.B., Vredenburgh W.A. *J. Org. Chem.*, 1966, 31, No.3, p.865-869.

[41] Voronkov M.G., Deryagina E.N. *Uspekhi khimii*, 1990, 59, No.12, p.1338-1361.

[42] Nanobashvili E.M., Chirakadze G.G., Movsumzade M.M., Radiation chemistry of sulfhydric compounds – Tbilisi: *Metznierba*, 1979, P.I, 84p.; 1980, P.II, 152p.

[43] Walling C., Helmreich W. *J. Amer. Chem. Soc.*, 1959, 81, No.4, p.1144-1148.

[44] Goldfarb Ya.L., Pokhil G.P., Belenkii L.I., *Zh. ob. khimii*, 1967, 37, No.11, p.2670-2676.

[45] Oswald A.A., Wallace T.J. *Anionic Oxidation of Thiols and Co-oxidation of Thiols with Olefins, in Organic Sulfur Compounds* (Ed. by N. Kharasch): – Pergamon Press, 1966, 2, Ch. 8, p.205-232.

[46] Griesbaum K., Oswald A.A., Hudson B.E. *J. Amer. Chem. Soc.*, 1963, 85, No.6, p.1969-1974.

[47] Wallace T.J., Schriesheim A. *J. Org. Chem.*, 1962, 27, No.5, p.1514-1516.

[48] Melnikov N.N., Novozhilov K.V., Belan S.R., Pylova T.N. Reference book on pesticides – Moscow: *Khimiya,* 1985, 352p.

[49] Yakhontov L.N., Vakhatova G.M. *Khim.-Farm. Zh.*, 1981, 15, No.8, p.27-41.

[50] Blokh G.A. Organic accelerators of rubbers vulcanization –Leningrad: *Khimiya*, 1972, 560p.

[51] Mamedov Sh., Mamedov A.A. Chemistry of sulfur organic compounds containing of oil and oil products: Collection of articles – Moscow: *Khimiya*, 1964, 6, p.208-210.

[52] Shostakovskii M.F., Prilezhaeva E.N., Uvarova N.I., Karavaeva V.M. Chemistry of sulfur organic compounds containing of oil and oil products: Collection of articles – Moscow: AN SSSSR, 1959, 2, p.337-350.

[53] Friedman M. *The Chemistry and Biochemistry of the Sulphydryl Group in Amino-acids, Peptides and Proteins* – Oxford: Pergamon Press, 1973, 877p.

[54] Lowe G. *Tetrahedron*, 1976, 32, No.3, p.291-302.

[55] Bikbulatov N.T. Chemistry of sulfur organic compounds containing of oil and oil products: Collection of articles – Moscow: *Vysshaya shkola*, 1972, 9, p.67-71.

[56] Shugaev B.B. Chemistry of sulfur organic compounds containing of oil and oil products: Collection of articles – Moscow: *Vysshaya shkola*, 1972, 9, p. 552-558.

[57] SabirovS.S., Gulin A.V., Pushkar N.V., *Khim.-Farm. Zh.*, 1985, 19, No.12, p.1453-1455.

[58] Rakhmankulov D.L., Kantor E.A., Zlotskii S.S. Synthesis and biological activity of 1,3-diheterocycloalkanes (subject review) – Moscow: *NIITEKhIM*, 1985, 24p.

[59] Liberman S.S., Yakhontov L.N. *Khim.-Farm. Zh.*, 1988, 22, No.9, p.1046-1058.

[60] Anichkov S.V., Sapronov N.S. Pharmacology of heart and vessels – Moscow: *Meditzina*, 1984, 204p.

[61] Leplyanin G.V., Rafikov S.R., Egforov Yu.I., et. al. *Vysokomol. Soed.*, 1976, 18B, No.1, p.22-215.

[62] Murinov Yu.I., Maistrenko V.N., Afzaletdinova N.G. Metals extraction by S, N-organic compounds – Moscow: *Nauka,* 1993, 190p.

[63] Murinov Yu.I., Khisamutdinov R.A., E.Yu. Nikitin, *Zh. neorg. khimii*, 1978, 23, No.11, p.3084-3088.

[64] Murinov Yu.I., Khisamutdinov R.A., E.Yu. Nikitin, *Zh. neorg. khimii*, 1980, 25, No.2, p. 500-503.

[65] Murinov Yu.I., Khisamutdinov R.A., Nikitin Yu.E. Hydrometallurgy of gold. Collection of articles – Moscow: *Metallurgiya*, 1980, p.143-148.

[66] Khan G.A., Gabrielova L.I., Vlasova N.S. Flotation reagents and their application – Moscow: *Nedra,* 1986, 276p.

[67] Koibash V.A., Koibash B.V. Chemistry of sulfurorganic compounds containing in oil and oil products. Collection of articles – Moscow: *Gostoptekhizdat,* 1961, 4, p.269-271.

[68] Abramov A.A. Technology of oxidized and mixed ores dressing – Moscow: *Nedra,* 1986, 512p.

[69] Makarova T.S., Perfilieva N.S., Kuzmichev G.V., et. al. XVI Conference on chemistry and technology of organic compounds of sulfur and sulfur oils – Riga, 1984, p.393-394.

[70] Pavelko G.F., Oleinik D.M., Bagrii E.I. *Neftekhimiya*, 1989, 29, No.4, p.547-550.
[71] Bolshakov G.F. *Neftekhimiya*, 1988, 28, No.2, p.238-246.
[72] Kuliev A.M. Chemistry and technology of additives for oils and fuels – Moscow: *Khimiya*, 1985, 312p.
[73] Kuliev A.B., Akhadov N.O., Abdullaeva M.I., Useinova G.G. *Neftekhimiya*, 1989, 29, No.4, p.536-540.
[74] Kuliev A.B., Abdullaeva M.I., Akhadov N.O. *Zh. prikl. khimii*, 1984, 57, No.8, p.1835-1838.
[75] Kuliev A.B., Akhadov N.O., Abdullaeva M.I., *Neftekhimiya*, 1987, 27, No.2, p.258-261.
[76] Kuliev A.B., Mamedov A.G., Dzhavadov M.M., et. al. *Neftekhimiya*, 1990, 30, No.6, p.841-845.
[77] Kuliev A.B., Dzhavadov M.M., Mamedov A.G. *Neftekhimiya*, 1990, 30, No.6, p. 846-849.
[78] Dogadkin B.A., Eitington I.I., Feldshtein M.S. *Colloid. zh.*, 1959, 21, No.4, p.427-435.
[79] Koval I.V. *Uspekhi khimii*, 1993, 62, No.8, p.813-830.
[80] Gavrilov L.E. Application of gas in national economy – Moscow: *VNIIEgasprom*, 1971, No.11, P. 17-21.
[81] Lebedev N.N. Chemistry and technology of basic organic and petrochemical synthesis – Moscow: *Khimiya*, 1985, 605p.
[82] Van Veen C.N.A., Moha-med K.A., Baller T., De Vries A.E. *J. Chem. Phys.*, 1983, 74, No.2, p.261-271.
[83] Mazaev V.E., Koprshunov M.A. *Catalysts of processes of reception and transformation of sulfur-containing compounds* – Novosibirsk: 1979, p.7-21.
[84] Mashkina A.V. *Uspekhi khimii*, 1995, 64, No.12, p.1210-1226.
[85] Korshunov M.A., BukharevaV.A., Furaeva I.V., Khokhlova L.V. Chemistry of sulfurorganic compounds containing in oil and oil products. Collection of articles – Moscow: *Vysshaya shkola*, 1968, 8, p.5-11.
[86] Ulendeeva A.D. Abstract of dissertation of doctor of science, Ufa: *IOC USC RAS*, 1998, 48p.
[87] Nikitina T.S., Lyapina N.K., Brodskii E.S., Ulendeeva A.D. *Neftekhimiya*, 1996, 36, No.2, p.105-115.
[88] Benkovskii V.G., Baikova A.Ya., Yarullin K.S. *Geologiya nefti i gasa*, 1973, No.10, p.35-42.
[89] Voltzov A.A., Lyapina N.K., Shmakov V.S., Parfenova M.A. *Neftekhimiya*, 1987, 27, No.5, p.616-624.
[90] Shmakov V.S., Ulendeeva A.D., Lyapina N.K., Furlei I.I. *Neftekhimiya*, 1989, 29, No.1, p.14-18.
[91] Lyapina N.K., Shmakov V.S., Parfenova M.A., Zinchenko L.I. *Neftekhimiya*, 1989, 29, No.2, p. 165-171.
[92] Lyapina N.K., Shmakov V.S., Parfenova M.A., *Neftekhimiya*, 1989, 29, No.4, p.453-457.
[93] Hopkins R.L., Smith H.M. *Anal. Chem.*, 1955, 27, No.12, p.1832-1833.
[94] Korolczuk J., Daniewcki M., Miciniczur Z. *J. Chromatogr.*, 1974, 100, No.1, p.165-168.

[95] Prokhorov G.M., Zainullina R.V., Lukmanova A.S., et. al. *Neftekhimiya*, 1976, 16, No.2, p.285-292.

[96] Khvostenko V.I. Mass-spectroscopy of negative ions on organic chemistry – Moscow: *Nauka,* 1981, 159p.

[97] Shmakov V.S., Furlei I.I., Lyapina N.K., et. al. *Dokl. AN SSSR*, 1986, 287, No.1, p.139-144.

[98] Shmakov V.S., Lyapina N.K., Furlei I.I., Tolstikov G.A. *Neftekhimiya*, 1988, 28, No.1, p.9-15.

[99] Shmakov V.S., Lyapina N.K., Furlei I.I., et. al. *Neftekhimiya*, 1988, 28, No.2, p. 277-283.

[100] Karaulova E.N. Chemistry of oil sulfides – Moscow: *Nauka*, 1970, 201p.

[101] Obolentzev R.D., Lyapina N.K., Ostanina T.M. Chemistry of sulfurorganic compounds containing in oil and oil products. Collection of articles – Moscow: *Vysshaya shkola*, 1968, 8, p. 394-397.

[102] Lyapina N.K., Ulendeeva A.D., Nikitina T.S., Nikitina V.S. *Chemistry and physics of oil and petrochemical synthesis* – Ufa: Bashkirskii filial AN SSSR, 1976, p. 219-223.

[103] M.A. Parfenova, Brodskii E.S., Zimina K.I. *Neftekhimiya*, 1975, 15, No.6, p.902-908.

[104] Lyapina N.K., Parfenova M.A., Nikitina V.S., et. al. *Neftekhimiya*, 1979, 19, No.6, p.921-927.

[105] Voltzov A.A., Lyapina N.K. *Khimiya i tekhnologiya topliv i masel*, 1980, No.4, p.37-40.

[106] Akopova G.S. Preparation and gas and gas condensate refinery – Moscow: *VNIIEGasprom*, 1978, No.5, p.17-23.

[107] Afanasiev A.I., Stryuchkov V.M., Podlegaev N.I., et. al. Preparation and gas and gas condensate refinery – Moscow: *VNIIEGasprom*, 1984, No.11, p.42.

[108] Lastovkina G.A., Radchenko E.D., Rudina M.G. Reference book for petroleum processor – Leningrad: *Khimiya*, 1986, 648p.

[109] Afanasiev A.I., Stryuchkov V.M., Podlegaev N.I. et. al Technology of refinery of sulfur-containing natural gas. Reference book. – Moscow: *Nedra*, 1993, 152p.

[110] Agaev G.A., Nasteka V.I., Seidov Z.D. Oxidative proceses of purification of sulfur-containing natural gases and hydrocarbon condensates – Moscow: *Nedra*, 1996, 301p.

[111] Nasteka V.I. Novel technologies of purification of high-sulfur natural gases and gas condensates – Moscow: *Nedra*, 1996, 107p.

[112] Nikolaev V.V., Busygina N.V., Busygin I.G. Main processes of physical and physical-chemical gas refinery – Moscow: OAO *Nedra*, 1998, 184p.

[113] Fridrikhsberg D.A. Course of colloidal chemistry – Leningrad: *Khimiya*, 1984, p.303.

[114] Stryuchkov V.M., Afanasiev A.I., Latyuk V.I., et. al. Preparation and gas and gas condensate refinery – Moscow: *VNIIEGasprom*, 1988, No.6, 30p.

[115] Bienko A.A., Berdnikov, Velmisov N.I., et. al. Technology of refinery of sulfur-containing natural gas and condensate – Orenburg, 1990, 197p.

[116] Mazgarov A.M., Vildanov A.F. *Neftekhimiya*, 1999, 39, No.5, p.371-378.

[117] Kuzmenko N.M., Afanasiev Yu.M., Frolov G.S., Glupanov V.N. Industrial and sanitary purification of gases (KhM-14) – Moscow: *TzINTIKhIM-NEFTEMASH*, 1987, 40p.

[118] Keltzev N.V. The basis of adsorption technics – Moscow: *Khimiya*, 1984, p. 383-404.

[119] Ione K.G. Polyfunctional catalysis of zeolites – Novosibirsk: *Nauka*, 1982, 272p.

[120] Shakhov A.D. Abstract of dissertation of candidate of science – Moscow: I.M. Gubkin *RSU of oil and gas*, 2001, 24p.

[121] Molchanov S.A. Abstract of dissertation of candidate of science – Moscow: I.M. *Gubkin RSU of oil and gas*, 2001, 24p.

[122] Obolentzev R.D. *Chemistry of sulfurorganic compounds containing in oil and oil products. Collection of articles – Ufa: Bashkirskii filial AN SSSR,* 1958, 1, p.187-200.

[123] Pronin V.A., Usoltzeva M.V., Trofimov B.A., et. al. *Zh. prikl. khimii*, 1973, 46, No.12, p.2602-2604.

[124] Bikbulatov N.T. Chemistry of sulfurorganic compounds containing in oil and oil products. *Collection of articles – Moscow: AN SSSR*, 1959, 2, p.369-374.

[125] Obolentzev R.D., Makova E.A., Kondratieva E.S., Prokhorov G.M. Chemistry of sulfurorganic compounds containing in oil and oil products. Collection of articles – Moscow-Leningrad: *Khimiya*, 1964, 7, p.24-30.

[126] Litvin O.B. The basis of rubbers synthesis technology – Moscow: Khimiya, 1972, 528p.

[127] Obolentzev R.D., Makova E.A., Vorobieva A.I., Ibragimov F.Sh. *Neftekhimiya*, 1970, 10, No.4, p.586-593.

[128] Obolentzev R.D., Makova E.A., Vorobieva A.I. Chemistry of sulfurorganic compounds containing in oil and oil products. Collection of articles – Moscow: *Vysshaya shkola*, 1972, 9, p.35-41.

[129] Bogomolov B.D. Chemistry of sulfurorganic compounds containing in oil and oil products. Collection of articles – Moscow: *Vysshaya shkola*, 1968, 8, p. 11-14.

[130] Shevchuk V.A., Leshinskii V.B., Tatevosyan L.K. Preparation and gas and gas condensate refinery – Moscow: *VNIIEGasprom*, 1980, No.3, p.10-14.

[131] Bobrova T.I., Volodkovich S.D., Kukalenko S.S. *Zh. org. khimii*, 1975, 45, No.8, p.1123-1125.

[132] Surface active agents. Reference book, ed. by Abramzon A.A., Gaevoi G.M. – Moscow: Khimiya, 1979, 376p.

[133] Ulendeeva A.D., Nasteka V.I., Latyuk V.I., et. al. *Neftekhimiya*, 1993, 33, No.6, p.547-551.

[134] Bugai D.E., Golubev M.V., Laptev A.B., et. al. *Bashkir. khim. zh.*, 1996, 3, No.4, p.59-63.

[135] Bolshakov G.F., Davydov P.I. Chemistry of sulfurorganic compounds containing in oil and oil products. Collection of articles – Moscow: *Gostoptekhizdat,* 1963, 5, p.138-148.

[136] Rubinshtain I.A., Sobolev E.P., Englin B.A., Danilova T.A. Chemistry of sulfurorganic compounds containing in oil and oil products. Collection of articles – Moscow: *Vysshaya shkola*, 1968, 8, p.492-508.

[137] Gureev A.A., Krivova T.I., Simanovskaya I.Ya. Chemistry of sulfurorganic compounds containing in oil and oil products. Collection of articles – Moscow: *Vysshaya shkola*, 1968, 8, p.542-546.

[138] Gureev A.A., Krivova T.I., Simanovskaya I.Ya. Chemistry of sulfurorganic compounds containing in oil and oil products. Collection of articles – Moscow: *Vysshaya shkola*, 1968, 8, p.509-515.

[139] Gureev A.A., Zarubina A.P., Subbotin A.P. Chemistry of sulfurorganic compounds containing in oil and oil products. Collection of articles – Moscow: *Vysshaya shkola*, 1968, 8, p.584-594.

[140] Shakhnovich M.I., Danilova A.I., Lipstain R.A., et. al. Chemistry of sulfurorganic compounds containing in oil and oil products. Collection of articles – Moscow: *Vysshaya shkola*, 1968, 8, p.538-542.

[141] Demchenko V.S., Novikov V.K. Chemistry of sulfurorganic compounds containing in oil and oil products. Collection of articles – Moscow: *Gostoptekhizdat*, 1961, 4, p. 194-198.

[142] Kelarev V.I., V.N. Koshelev, Golubeva I.A., Malova O.V. Stabilizers and modifiers of organic materials on the base of *sym*-triazine derivatives – Moscow: *TzNIITEneftekhim*, 1996, 64p.

[143] Kelarev V.I., Silin M.I., Koshelev V.N. Perspective processes and products of small-tonnage chemistry. Collection of reviews – Ufa: *Reaktiv*, 2000, No.4, p. 43-55.

[144] Grunvald V.R., Galperin B.M., Lavrenko A.I., Fakhriev A.M. *Gazovaya promyshlennost*, 1983, No.2, p.32-33.

[145] Laptev N.G., Bogoslovskii V.M. Chemistry of dyes – Moscow: *Khimiya*, 1970, 423p.

[146] Malyatova G.P., Chernomyrdina N.A., Kapralova N.N., Grunvald V.R. *Gazovaya promyshlennost*, 1983, No.5, p.36-37.

[147] Arshid F.M., Giles G.H., McLure E.C., et. al. *J. Chem. Soc.*, 1955, No.1, p.67-79.

[148] Kolychev V.B., Paramonova V.I. *Spectroscopic methods in chemistry of complex compounds*, ed. by V.M. Vdovenko – Moscow;Leningrad: 1964, p.30-73.

[149] Bystrov V.F. Hydrogen bond. Collection of articles – Moscow: *Nauka*, 1964, p. 253-266.

[150] Leshev S.M., Melsitova I.V. *Neftekhimiya*, 1993, 33, No.2, p.186-190.

[151] Author certificate 1027156 Russia 07.07.1983.

[152] Author certificate 1077884 Russia 07.03.1984.

[153] Patent 1416159 Russia 1988.

[154] Patent 1125230 Russia 1984.

[155] Patent 1214730 Russia 1986.

[156] Patent 2105597 Russia 1998.

[157] Truce W.E., Hollister K.R., Lindy L.B., Parr J.E. *J. Org. Chem.*, 1968, 33, No.1, p.43-48.

[158] Bird R., Stirling C.J.M *J. Chem. Soc. B.*, 1968, No.1, p.111-115.

[159] Shostakovskii S.M., Bobrov A.V. *Zh. org. khimii*, 1969, 5, No.6, p.908-914.

[160] Derzhinskii A.R., Chizhov O.S., Prilezhaeva E.N. *Izv. AN SSSR. Ser. Khim.*, 1977, No.4, p.850-854.

[161] Derzhinskii A.R., Kochetkova N.K., Nekhoroshev S.A., Piottukh-Peletzkii V.N. *Izv. AN SSSR. Ser. Khim.*, 1981, No.11, p.2597-2605.

[162] Patent 3249573 USA 1966.

[163] Derzhinskii A.R., Konyushkin L.D., Prilezhaeva E.N. *Izv. AN SSSR. Ser. Khim.*, 1978, No.9, p.2070-2076.

[164] Derzhinskii A.R., Konyushkin L.D., Prilezhaeva E.N. *Izv. AN SSSR. Ser. Khim.*, 1982, No.5, p. 1116-1123.

[165] Lindner K. Tenside, Textilhilfsmittel, Waschrochstoffe – Stuttgart: *Wisseschaftliche Verlagsgesellschaft,* 1964, 1241s.
[166] Daier J.R. Applications to absorption spectroscopy of organic compounds, transl. from English – Moscow: *Khimiya,* 1970, 163p.
[167] Silverstain R., Basler G., Morril T. Spectroscopic identification of organic compounds, transl. from English – Moscow: *Mir,* 1977, 590p.
[168] General organic chemistry, ed. by D.Barton, Y.D. Wollis, transl. from English – Moscow: *Khimiya,* 1983, 728p.
[169] Brook M.A., Chan T.H. *Synthesis*, 1983, No.3, p.201-203.
[170] Chandrasekaran S., Turner J.V. *Synth. Commun.*, 1982, 12, No.9, p.727-730.
[171] Carsia T., Arrieta A., Palomo C. *Synth. Commun.*, 1982, 12, No.9, p.681-684.
[172] K.I. Kobrakov, V.I. Kelarev, I.M. Rybina, et. al. *Bash. khim. zh.*, 2001, 8, No.4, p.3-16.
[173] Heterocyclic compounds. *Collection of articles*, ed. by R. Elderfild, transl. from English – Moscow: IL, 1961, 602p.
[174] Kelarev V.I., Koshelev V.N. *Uspekhi khimii*, 1995, 64, No.4, p.339-372.
[175] Kelarev V.I., Shvekhgeimer G.A. *Izv. VUZov. Khimiya i khim. tekhnologiya*, 1981, 24, No.11, p.1354-1358.
[176] Koshelev V.N., Kelarev V.I., Lunin A.F. *Zh. D.I. Mendeleev VKhO*, 1984, 29, No.4, p.470-472.
[177] Kelarev V.I., Karakhanov R.A., Gasanov S.Sh. *Khimiya geter. soed.*, 1993, No.2, p.189-196.
[178] Nesynov E.P., Grekov A.P. *Uspekhi khimii*, 1964, 33, No.10, p.1184-1205.
[179] Kelarev V.I., Koshelev V.N., Karakhanov R.A. *Bash. khim. zh.*, 1996, 3, No.1-2, p.79-83.
[180] Silin M.A., Kelarev V.I., Gracheva O.G., et. al. *Bash. khim. zh.*, 1998, 5, No.2, p.14-17.
[181] Karakhanov R.A., Kelarev V.I., Remizov A.S. *Khimiya geter. soed.*, 1992, No.10, p.1312-1315.
[182] Piskov V.B., Kasperobich V.P. *Zh. org. khimii*, 1978, 14, No.4, p.820-825.
[183] Physical methods in chemistry of heterocyclic compounds. Collection of articles, ed. by Katritzkogo A.R. – Moscow-Leningrad: *Khimiya,* 1966, 628p.
[184] Sasaki T., Eguchi S., Toru T. *Bull. Soc. Chim. Japan.*, 1969, 42, No.6, p.1617-1620.
[185] Brown H.C., Pilipovich D. *J. Am. Chem Soc.*, 1960, 82, No.16, p.4700-4703.
[186] Dziewonska M. *Spectrochim. Acta*, 1967, 23A, No.8, p.1195-1204.
[187] Karakhanov R.A., Kelarev V.I., Koshelev V.N. *Khimiya geter. soed.*, 1995, No.2, p.238-244.
[188] Imai Y., Taoka P., Uno K., Iwakura Y. *Makromolek. Chem.*, 1965, 83, No.2, p.167-178.
[189] Hayashi S., Furukawa M., Fujino Y., Yoshimatsu S. *Chem. Pharm. Bull.*, 1969, 17, No.2, p.329-334.
[190] Shapiro S.L., Isaaks E.S., Freedman L. Guanamines. V. *J. Org. Chem.*, 1961, 26, No.1, p.74-76.
[191] Zakharenko V.A. Herbicides – Moscow: *Agropromizdat*, 1990, 264p.
[192] Pogosyan G.M., Pankratov V.A., Zaplishnyi V.N., Matzoyan S.G. Polytriazines – Erevan: *Izd-vo AN ArmSSR*, 1987, 615p.
[193] Mur V.N. *Uspekhi khimii*, 1964, 33, No.2, p.182-204.

[194] *Stabilizers, accelerators, vulcanization agents, porophores abroad* (catalogue), ed. by Burmistrov V.F. – Tambov, 1972, 359p.

[195] Kasparov V.A., Promonenkov V.K. Application of pesticides abroad – Moscow: *Agropromizdat,* 1990, 215p.

[196] Davis D.W. *Trans. Faraday Soc.*, 1955, 51, No.4, p.449-456.

[197] Koopman H., Daams J. *Rec. Trav. Chim.*, 1960, 79, No.1, p.83-89.

[198] Patent 1176770 England 1970.

[199] Eremin S.A., Samsonova Zh.V., Egorov A.M. *Uspekhi khimii*, 1994, 63, No.7, p.638-649.

[200] Malova O.V., Vishnyakova T.P., Golubeva I.A., et. al. *Khimiya geter. soed.*, 1984, No.12, p.1678-1682.

[201] Kelarev V.I., Belov N.V., Remizov A.S., et. al. *Izv. VUZov. Khimiya i khim. tekhnologiya*, 1994, 37, No.10-12, p.16-23.

[202] Kelarev V.I., Morozova G.V., Koshelev V.N., et. al. *Izv. VUZov. Khimiya i khim. tekhnologiya*, 1997, 40, No.3, p.83-89.

[203] Kelarev V.I., Koshelev V.N., Silin M.A. *Izv. VUZov. Khimiya i khim. tekhnologiya*, 1998, 41, No.2, p.14-20.

[204] Silin M.A., Kelarev V.I., Abu-Ammar V.M., et. al. *Neftekhimiya*, 2000, 40, No.3, p.235-240.

[205] Ershov V.V., Nikiforov G.A., Volodkin A.A. Spatially hindered phenols – Moscow: *Khimiya,* 1972, 352p.

[206] Ishonova Kh. Dissertation of candidate of science, V.I. Nikitin Institute of chemistry *AN TadzhSSR,* 1986, 139p.

[207] Paken A.M. Epoxide compounds and epoxide resins – Leningrad: Goskhimizdat, 1962, 963p.

[208] Kuliev A.M., Farzaliev V.M., Alizade Z.A., Alakhverdiev M.A. Organic compounds of sulfur. *Collection of articles – Riga*: *Zinatne*, 1980, 2, p.18-27.

[209] Bellami L. *Infra-red spectra of complex molecules* – Moscow: IL, 1966, 346p.

[210] Zaikin V.G., Varlamov A.V., Mikaya A.I., Prostakov N.S. The basis of mass-spectrometry of organic compounds – Moscow: *MAIK Nauka*, 2001, 283p.

[211] Nakanisi K. Infra-red spectra and structure of organic compounds – Moscow: *Mir*, 1965, 214p.

[212] Rakhmankulov D.L., Zorin V.V., Latypova F.N. Methods of synthesis of 1,3-heteroanalogues of cycloalkanes – Ufa: *Reaktiv*, 1998, 248p.

[213] Rakhmankulov D.L., Zlottzkii S.S., Antonova A.M., Fatkullina F.A. IR-spectra of some heterorganic compounds – Ufa: *Izd-vo UNI,* 1989, 80p.

[214] Bergman E., Ginzburg D., Pappo R. Organic reactions. *Collection of articles – Moscow: IL,* 1963, 10, p.181-253.

[215] Butzkus P.F., Denis G.I. *Uspekhi khimii*, 1966, 35, No.11, p.1999-2019.

[216] Zilberman E.N. Nitriles reactions – Moscow: *Khimiya,* 1972, 384p.

[217] Tzoi L.A., Sadykov B.T., Sokolov D.V., Omarov T.T. *Khimiya geter. soed.*, 1968, No.3, p.549-553.

[218] Kelarev V.I., Karakhanov R.A., Morozova G.V., et. al. *Khimiya geter. soed.*, 1994, No.1, p.115-118.

[219] Goldin G.S., Maksakova M.V., Poddubnyi V.G., Koltzova A.N. *Zh. obsh. khimii*, 1975, 45, No.12, p.2661-2669.
[220] Grekov A.P. Organic chemistry of hydrazine – Kiev: *Tekhnika,* 1996, 235p.
[221] Overberger J., Anselm Zh.-P., Lombardino D.G. Organic compounds with bonds nitrogen–nitrogen, transl. from English – Leningrad: *Khimiya*, 1970, 364p.
[222] Kitaev Yu.P., Buzykin B.I. Hydrazones – Moscow: *Nauka*, 1974, 405p.
[223] Chemistry of hydrazones. Collection of articles, ed. by Kitaev Yu.P. – Moscow: *Nauka*, 1977, 388p.
[224] Jensen K.A., Nielsen P.H. *Acta Chem. Scand.*, 1966, 20, No.4, p.597-629.
[225] Polivin Yu.N., Karakhanov R.A., Kelarev V.I., Bienko A.A. *Izv. VUZov. Khimiya i khim. tekhnologiya*, 1992, 35, No.10, p.31-34.
[226] Kelarev V.I. Dissertation of doctor of science – Moscow: I.M. Gubkin GANG, 1992.
[227] General organic chemistry, ed. by D.Barton, Y.D. Wollis, transl. from English – Moscow: Khimiya, 1985, 8, 752p.
[228] Sheltzin V.K. Synthesis of benzimidazoles – Moscow: *NIITEKhIM*, 1984, 64p.
[229] Krasovitzkii B.M., Bolotin B.M. Organic luminophore – Moscow: *Khimiya*, 1984, 336p.
[230] *Thiazole and Its Derivatives*, ed. by J.V.Metzger – New York: John Willey a. Sons, Part 1, 1979, 672 p.
[231] G.R.Rao, M.T.Chary, K.Mogilaiah et al. *J. Indian Chem. Soc.*, 1989, 66, No.1, p.61-63
[232] Fernandes P.S., Sonar T.M. *J. Indian Chem. Soc.*, 1988, 65, No.1, p.46-48.
[233] Yashunskii V.G., Kobtun G.Yu. *Uspekhi khimii*, 1985, 54, No.1, p.126-161.
[234] Kelarev V.I., Silin M.A., Abu-Ammar V.M., Golubeva I.A. *Neftekhimiya*, 2000, 40, No.1, p.53-57.
[235] Heterocyclic compounds, ed. by Elderfild, transl. from English – Moscow: *Mir*, 1965, 7, p. 296-328.
[236] Prameela B., Rajaharendar E., Murty A.K. *Collect. Czech. Chem. Commun.*, 1989, 54, No.7, p.1092-1095.
[237] Preston P.N. *Chem. Rev.*, 1974, 74, No.3, p.279-314.
[238] Braz G.N., Myasnikova G.V., Yakubovich A.Ya., et. al. *Khimiya geter. soed.*, 1967, No.2, p.215-219.
[239] Kelarev V.I., Shvekhgeimer G.A. *Khimiya geter. soed.*, 1982, No.3, p.343-347.
[240] Kelarev V.I., Shvekhgeimer G.A. *Izv. VUZov. Khimiya i khim. tekhnologiya*, 1982, 26, No.12, p.1458-1461.
[241] Grekov A.P., Solovieva M.S. *Zh. obsh. khimii*, 1960, 30, No.7, p.1614-1650.
[242] Massay D.J.R. *Synthesis*, 1987, No.7, p.589-591.
[243] Silin M.A., Kelarev V.I., Abu-Ammar V.M. *Khimiya geter. soed.*, 2000, No.2, p.256-260.
[244] Kelarev V.I., Shvekhgeimer G.A., Lunin A.F. *Khimiya geter. soed.*, 1984, No.9, p.1271-1274.
[245] Kelarev V.I., Snegotzkii V.A., Kobrakov K.I., et. al. *Izv. TulGU Ser. Khim.*, 2003, No.4, p.67-79.
[246] Skvortzova G.G., Glazkova N.P., Domina E.S., Voronov V.K. *Khimiya geter. soed.*, 1970, No.2, p.167-171.
[247] Mikhaleva A.I., Korostova E.S., Vasiliev A.N., et. al. *Khimiya geter. soed.*, 1977, No.12, p.1636-1639.

[248] Dronov V.I., Nikitin Yu.E. *Uspekhi khimii*, 1985, 54, No.6, p.941-955.
[249] Blik F.F. Organic reactions. *Collection of articles – Moscow: IL*, 1948, 1, p.399-454.
[250] Hellmann H., Opitz G. α-Aminoalkylierung – Weinheim: *Verlag-Chemie*, 1960, 588s.
[251] Mathien J., Weill-Raunal J. Introduction of One Functional Carbon Atom – Stuttgart: *Georg. Thiele Publ.*, 1973, p.56-81.
[252] Tramontini M. *Synthesis*, 1973, No.12, p.703-707.
[253] Pokonova Yu.V. Haloidsulfides – Leningrad: *Izd-vo LGU,* 1977, 278p.
[254] Böhme H., Bentler H. *Chem. Berichte*, 1956, 89, No.6, s.1464-1468.
[255] Tuleen D.L., Stephens T.B. *J. Org. Chem.*, 1969, 34, No.1, p.31-35.
[256] Turchaninova L.P., N.A. Korchevin, Shipov A.G., et. al. *Zh. obsh. khimii*, 1989, 59, No.3, p.722-723.
[257] Shukurov S.Sh., Negmatullaev Kh.N., Daniyarova M.S., Sabirob S.S. *Zh. org. khimii*, 1979, 15, No.5, p.1102-1103.
[258] Minkin V.Yu., Zhdanov Yu.A., Medyantzeva E.A. Azomethines – Rostov-on-Don: *Izd-vo RostGU,* 1967, 232p.
[259] Dhar D.N., Sharma R.L. *Indian J. Anim. Sci.*, 1988, 58, No.10, p.1150-1153.
[260] Kuznetzov V.V., Prostakov N.S. *Khimiya geter. soed.*, 1990, No.1, p.5-32.
[261] Byuler K., Pirson D. Organic synthesizes, transl. from English – Moscow: *Mir*, 1973, 2, p. 483-486.
[262] Veigand K., Khilgetag G. Methods of experiment in organic chemistry – Moscow: *Khimiya*, 1968, p. 518-519.
[263] Khaiosh A. Complex hydrides in organic chemistry, transl. from Deutsch, ed. by N.N. Suvorov – Leningrad: *Khimiya*, 1971, 624p.
[264] Reninskaya I.B., Shvartzberg M.S. Selection of organic compounds synthesis methods – Novosibirsk: *Izd-vo Novosibirskogo universiteta*, 2000, 284p.
[265] Modern methods of organic synthesis. Collection of articles, ed. by Ioffe B.V. – Leningrad: *Izd-vo Leningradskogo universiteta*, 1980, 232p.
[266] Rakhmankulov D.L., Karakhanov R.A., Zlottzkii S.S., et. al. *Itogi nauki. Ser. Tekhn. org. veshestv*, Moscow: *VINITI*, 1979, 5, 288p.
[267] Rakhmankulov D.L., Zorin V.V., Latypova F.N., et. al. *Uspekhi khimii*, 1983, 52, No.4, p.619-631.
[268] Khlebnikova T.D., Pokalo E.I., Tolmacheva N.A., et. al. Perspective processes and products of small-tonnage chemistry. *Collection of articles – Ufa: Reaktiv,* 2000, No.4, p.68-82.
[269] Khabibulin I.R., Rolnik L.Z., Zlottzkii S.S., Rakhmankulov D.L., *Zh. prikl. khimii*, 1991, 64, No.2, p.443-447.
[270] Rolnik L.Z., Yagafarova G.G. Perspective processes and products of small-tonnage chemistry. *Collection of articles – Ufa: Reaktiv,* 2000, No.4, p.120-126.
[271] Kantor E.A., Imasheva U.B., Musavirov R.S., et. al. Ufa: *Izd-vo UNI,* 1980, 93p.
[272] Harris T.D., Roth G.P. *J. Org. Chem.*, 1979, 44, No.12, p.2004-2008.
[273] Carpenter A.J., Chadwick D.J. *Tetrahedron*, 1985, 41, No.18, p.3803-3812.
[274] Хабибулин С.Г., Вольцов А.А. Современные методы извлечения сернистых соединений из нефтей // Сернистые нефти и продукты их переработки: сб. статей. – Уфа: *Баштехинформ*, 1994. – вып. 22. – С. 80-91.

[275] Вольцов А.А., Исмагилов Ф.Р., Вольцов Ан.А. Экологически безопасная глубокая переработка газовых конденсатов // *Химия и технология топлив и масел*. – 1999. - No. 4. – С. 3-5.

[276] Борисенкова С.А., Вильданов А.Р., Мазгаров А.М. Современные проблемы обессеривания нефтей и нефтепродуктов // *Рос. хим. ж*. – 1995. – Т. 39. - No. 5. – С. 87-101.

[277] Рубинштейн И.А., Клеменова З.А., Соболев Е.П. Титриметрические методы определения меркаптановой и сульфидной серы в нефтях и нефтепродуктах // Методы анализа органических соединений нефти, их смесей и производных: сб. статей. – М.: *Изд. АН СССР*, 1960. – Т. 1. – С. 74-78.

[278] Дронов В.И., Нигматулина Р.Ф., Никитин Ю.Е. Реакция тиоалкилирования. II. Конденсация кетонов с формальдегидом и 2-метил-2-пропан-, циклогексантиолами или тиофенолом // *Ж. орг. химии*. – 1977. – Т. 13. - No. 10. – С. 1604-1610.

[279] Оболенцев Р.Д., Ляпина Н.К., Никитина В.С., Парфенова М.А. Изучение экстракции сераорганических соединений серной кислотой из нефтяных дистиллятов // *Нефтехимия*. – 1971. – Т. 11. - No. 1. – С. 125-129.

[280] Parham W.E., Ramp E.L. β-Nitrosulfides and aminosulfides // *J. Am. Chem. Soc*. – 1951. – V. 73. - No. 9. -P. 1293-1297.

[281] Козлова Е.А., Котова Г.Г., Фукс Г.И., Эминов Е.А. О механизме синергетического действия "парных антиокислителей" // *Химия и технол топлив и масел*. - 1982. - No. 2. – С. 31-32.

[282] Вилянская Е.Д., Куликовская Т.Н., Знаменская О.А., Бутягина Е.В. О механизме действия парной антиокислительной присадки //*Химия и технология топлив и масел*. – 1988. - No. 8. – С. 36-38.

[283] Влияние композиционных присадок, содержащих аминопроизводные сим-триазина, на термоокислительную стабильность минеральных смазочных масел / В.И.Келарев, В.Н.Кошелев, О.Г.Грачева и др. // *Изв. вузов. Нефть и газ*. – 1999. - No. 1. - С. 97-103.

[284] Топлива. Смазочные материалы. Технические жидкости: Справочник / Под ред. В.М.Школьникова. М.: *Химия*, 1999. - 596 с.

[285] *Химия смазочных материалов и топлив: сб. лабораторных работ* (Ч. 2) / Под ред. Л.П.Казаковой. – М.: ГАНГ им. И.М.Губкина, 1988. – С. 45-47.

[286] Гафаров Н.А., Гончаров А.А., Кушнаренко В.М. Коррозия и защита оборудования сероводородсодержащих нефтегазовых месторождений. – М.: *Недра*, 1998. – 437 с.

[287] Спиркин В.Г., Бочаров А.А. Надежность и ресурсы работы газовых компрессоров // *Газовая пром-ть*. – 1996. - No. 5. – С. 50-51.

[288] Спиркин В.Г., Гильмутдинов Ш.К., Ахметшин Э.А. Оценка смазывающих свойств масел в присутствии сероводорода // *Химия и технология топлив и масел*. – 1993. - No. 6. – С. 31-32.

[289] Горбунов Б.Н., Гурвич Я.А., Маслова И.П. Химия и технология стабилизаторов полимерных материалов. – М.: *Химия*, 1981. – 368 с.

[290] Денисов Е.Т., Ковалев Г.И. Окисление и стабилизация реактивных топлив. – М.: *Химия*, 1983. – 269 с.

[291] Вишнякова Т.П., Голубева И.А., Крылов И.Ф., Лыков О.П. Стабилизаторы и модификаторы нефтяных дистиллятных топлив. – М.: *Химия,* 1990. – 191 с.

[292] Энглин Б.А., Борисова С.М. Изучение стабилизации реактивных топлив присадками // *Химия и технол. топлив и масел*. – 1972. - No. 8. – С. 15-17.

[293] Гуреев А.А., Серегин Е.П., Азев В.С. Квалификационные методы испытаний нефтяных топлив. – М.: *Химия,* 1984. – 200 с.

[294] Чертков Я.Б. *Моторные топлива*. – Новосибирск, 1987. – 208 с.

[295] Данилов А.М. Присадки и добавки. Улучшение экологических характеристик нефтяных топлив. – М.: *Химия,* 1996. – 231 с.

[296] Данилов А.М. Проблемы окислительной стабильности вторичных дистиллятных топлив // *Нефтехимия*. – 1992. – Т. 32. - No. 3. – С. 374-376.

[297] Курамшин Э.М.,Имашев У.Б. Окисление и стабилизация дизельных топлив (Тем. обзор. – сер.: Переработка нефти). – М.: *ЦНИИТЭнефтехим*, 1995. – 141 с.

[298] Шукуров С.Ш. Синтез, химические превращения и биологическая активность производных 1,3,4-тиадиазола и 1,3,4-тиадиазоло(3,2-а)пиримидина: Дис….д.х.н. (науч. докл.) – М.: ГАНГ им. И.М.Губкина, 1993. – 60 с.

[299] Джалолов С.С. Синтез и химические превращения галогенпроизводных 1-тиаиндана: Дис….д.х.н. (науч. докл.) – М.: ГАНГ им. И.М.Губкина, 1993. – 59 с.

[300] Куатбеков А.М. Синтез, химические превращения и биологическая активность некоторых производных азолов, диазолов, пиперидина и сим-триазина: Автореф. дис….д.х.н. – М.: ГАНГ им. И.М.Губкина, 1994. – 47 с.

[301] Крапивин Г.Д. Фурансодержащие гетероциклические системы: синтез, стереостроение, реакционная способность и биологическое действие: Автореф. дис. … д.х.н. – Саратов: Саратовский ГУ им. Н.Г.Чернышевского, 1995. – 53 с.

[302] Кошелев В.Н. Синтез и свойства пяти- и шестичленных азотсодержащих гетероциклических соединений с двумя и тремя гетероатомами: Автореф. дис….д.х.н. – М.: ГАНГ им. И.М.Губкина, 1996. – 53 с.

[303] Котов А.Д. Синтез и свойства производных 3-арил-2,1-бензизоксазола: Автореф. дис….д.х.н. – М.: МГТУ им. А.Н.Косыгина, 2001. – 50 с.

[304] Богза С.Л. Разработка новых методов синтеза полядерных карбо- и гетероциклических соединений на основе кислотно-катализируемых реакций: Автореф. дис. … д.х.н. – М.: МГТУ им. А.Н.Косыгина, 2003. – 32 с.

[305] Капо-Чичи К. Разработка методов рациональной утилизации фракции меркаптанов, выделяемой из газового конденсата оренбургского месторождения: Автореф. дис. … к.х.н. – М.: ГАНГ им. И.М.Губкина, 1994. – 23 с.

[306] Белов Н.В. Синтез и химические превращения тиопроизводных сим-триазина: Автореф. дис. … к.х.н. – М.: ГАНГ им. И.М.Губкина, 1996. – 26 с.

[307] Абу-Аммар В.М. Производные пространственно-затрудненных фенолов с CH=N-группировками. Синтез, свойства и применение: Автореф. дис. … к.х.н. – М.: РГУ нефти и газа им. И.М.Губкина, 2000. – 25 с.

[308] Григорьева Н.А. Синтез и свойства функциональных производных бензтиазолил-2-тиоуксусной кислоты: Автореф. дис. … к.х.н. – М.: РГУ нефти и газа им. И.М.Губкина, 2000. – 27 с.

[309] Власова И.Д., Вишнякова Т.П., Гречушкина Н.Н. Биологическое поражение нефти и нефтепродуктов и их защита при транспортировке и хранении (Тем. обзор. – сер.: Переработка нефти). – М.: *ЦНИИТЭнефтехим*, 1970. – 55 с.

[310] Флеров Б.К. Проблемы биологических повреждений и обрастания материалов. – М.: *Наука*, 1972. – 410 с.

[311] Алкилфеноксистаннаны как биоцидные присадки к смазочным материалам / П.С.Белов, О.Н.Цветков, Н.Н.Комарова и др. // *Химия и технология топлив и масел*. – 1980. - No. 4. – С. 43-46.

[312] Биоцидная активность 1,3-дигетероцикланов, содержащих фрагмент пространственно-затрудненного алкилфенола / П.Ш.Мамедова, К.И.Садыхова, Э.Р.Баба-ев и др. // Перспективные процессы и продукты малотоннажной химии: *сб. статей. – Уфа: Реактив*, 1998. – С. 56-59.

[313] Единая система защиты от старения и коррозии масел и смазок: ГОСТ 9.048-75 – 9.052.75. – М.: Госстандарт, 1975. – 68 с.

[314] Ибрагимов Л.Х., Мищенко И.Т. Интенсификация добычи нефти. – М.: *Нефть и газ,* 1996. – 478 с.

[315] Микробиологическое воздействие на нефтяной пласт / Могилевский Г.А., Оборин А.А., Королев В.К. и др. – М.: ВНИИОЭНГ, 1979. – сер.: *Нефтепромысло-вое дело. – обз. информ.* – 44 с.

[316] Химические реагенты в добыче и транспорте нефти: Справочник / Д.Л.Рахманкулов, С.С.Злотский, В.И.Мархасин и др. – М.: *Химия*, 1987. – 144 с.

[317] Ибрагимов Г.З., Сорокин В.А., Хисамутдинов Н.И. Химические реагенты для добычи нефти. - М.: *Недра*, 1986. – 240 с.

[318] Авт. свид. 981244 СССР. МКИ С 02 F 1/50. Способ подавления жизнедеятельности сульфатвосстанавливающих бактерий / Л.Г.Шакиров, Р.Х.Хазипов, А.З.Биккулов и др. – заявл. 09.03.81, опубл. 15.12.82. – *РЖХим*. – 1983. – 17П112П.

[319] Авт.свид. 1573002 РФ. МКИ С 02 F 1/50. Реагент для подавления сульфат-редукции в заводняемом нефтяном пласте / М.М.Люшин, Ф.Ф.Муганлинский, А.Г.Ханларова и др. – заявл. 13.12.85, опубл. 23.06.90. – *РЖХим*. – 1990. – 21П140П.

[320] Авт. свид. 1570999 РФ. МКИ С 02 F 1/50. Реагент для подавления сульфат-редукции в заводняемом нефтяном пласте / М.М.Люшин, Ф.Ф.Муганлинский, А.Г.Ханларова и др. – заявл. 18.04.86, опубл. 15.06.90. – *РЖХим*. – 1990. – 22Н146П.

[321] Милованова С.И. Лекарственная регуляция воспалительных процессов // Методы экспериментальной химиотерапии: сб. статей. – М.: *Медгиз*, 1971. – С. 100-106.

[322] Першин Г.Н. Методика оценки острой токсичности биологически активных веществ // Методы экспериментальной химиотерапии: сб. статей. – М.: *Медгиз*, 1971. – С. 107-115.

[323] Гацура В.В. Методы первичного фармакологического исследования биологически активных веществ. – М.: *Медгиз,* 1974. – С. 27-34.

[324] Биоантиоксиданты в лучевом поражении и злокачественном росте / Е.Б.Бур-лакова, А.В.Алесенко, Е.М.Молочкина и др. – М.: *Наука*, 1975. – 211 с.

[325] Тиунов Л.А., Васильев Г.А., Вальдштейн Э.А. Противолучевые средства. – М.-Л.: *Медгиз,* 1964. – 325 с.

[326] Беленький М.Л. Элементы количественной оценки фармакологического эффек-та. – Л.: *Наука*, 1963. – 252 с.

[327] Миллер Д.Ж. Эксперименты в молекулярной генетике / Пер. с англ. – М.: *Мир*, 1976. – 368 с.
[328] Иоффе Б.В. Рефрактометрические методы в химии. – М.: *Химия*, 1983. – 224 с.

INDEX

C

D

E

F

G

H

I

P

Q

R

T

U

V

W

Y

Z